Introduction to Recognition and Deciphering of Patterns

Introduction to Recognition and Deciphering of Patterns

Michael A. Radin

School of Mathematical Sciences
Rochester Institute of Technology
Rochester, New York

CRC Press
Taylor & Francis Group
Boca Raton London New York

CRC Press is an imprint of the
Taylor & Francis Group, an **informa** business

A CHAPMAN & HALL BOOK

CRC Press
Taylor & Francis Group
6000 Broken Sound Parkway NW, Suite 300
Boca Raton, FL 33487-2742

International Standard Book Number-13: 9780367407278 (Hardback)
International Standard Book Number-13: 9780367508609 (Paperback)

Library of Congress Cataloging-in-Publication Data

Names: Radin, Michael A. (Michael Alexander), author.
Title: Introduction to recognition and deciphering of patterns / Michael A. Radin.
Description: First edition. | Boca Raton, FL : CRC Press, 2020. | Includes bibliographical references and index.
Identifiers: LCCN 2020015534 (print) | LCCN 2020015535 (ebook) | ISBN 9780367407278 (hardback) | ISBN 9780367508609 (paperback) | ISBN 9780367808747 (ebook)
Subjects: LCSH: Pattern perception. | Recurrent sequences (Mathematics) | Recognition (Psychology)
Classification: LCC BF294 .R344 2020 (print) | LCC BF294 (ebook) | DDC 152.14/23--dc23
LC record available at https://lccn.loc.gov/2020015534
LC ebook record available at https://lccn.loc.gov/2020015535

Visit the Taylor & Francis Web site at
http://www.taylorandfrancis.com

and the CRC Press Web site at
http://www.crcpress.com

Contents

Preface

VARIOUS PATTERNS ARE FREQUENTLY USED IN MANY ARTISTIC decorations and in architectural designs. Recognizing patterns and studying their unique traits is essential for the development and enhancement of our intuition, analytical skills, and inductive and deductive reasoning. It is vital to decipher distinct patterns when studying the weather systems, when learning to play a musical instrument, when studying a foreign language, when learning computer programming, when studying traffic patterns and transportation schedules, when analyzing engineering structures, when processing signals, when studying behaviors, when studying economic cycles and additional similar applications.

The aim of this interdisciplinary book is to gain practical experience in recognizing the emergence of specific patterns while studying complex structures of geometrical figures such as system of squares, systems of triangles, system of circles and additional geometrical systems and formations; this will direct us to exploring repetitions of patterns at same scale, repetition of patterns at different scales and alternating patterns. These analyses will guide us to the study of describing peculiar sequences, summations and products: linear sequences, quadratic sequences, summation-type sequences, geometric sequences, factorial-type sequences, piecewise sequences, recursive sequences, periodic sequences and specific patterns of Pascal's triangle.

I invite you on a voyage of discovery in recognizing and deciphering various patterns analytically and geometrically. The immense study of patterns will help us strengthen our intuitive skills and will guide us to an inductive approach and cognizance of other branches of mathematics and STEM-related disciplines. For each topic we will provide several step-by-step repetitive-type examples that will lead to inductively expanding our knowledge and comprehension about assorted patterns and their applications. During the last six years, I successfully implemented a hands-on teaching style with repetitive-type examples in the courses that I teach at the Rochester Institute of Technology and in the courses and seminars that I teach annually in Latvia at Liepaja University, at Riga Technical University and at Rezekne Technical Academy. It is a very rewarding experience to receive positive and supportive students' and colleagues' feedback and evaluations.

Michael A. Radin

Author

Michael A. Radin earned his PhD at the University of Rhode Island in 2001 and is currently an associate professor of mathematics at the Rochester Institute of Technology. He started his journey analyzing difference equations that portray periodic and eventually periodic solutions as part of his PhD thesis and has several publications on boundedness and periodic nature of solutions of rational difference equations, max-type difference equations and piecewise difference equations. Michael published several papers together with his master's students and undergraduate students at RIT and has publications with students and colleagues from Riga Technical University and the University of Latvia.

Michael also has publications in applied mathematics and related topics such as Neural Networking, Modeling Extinct Civilizations, and Modeling Human Emotions. In addition, he organized numerous sessions on difference equations and applications at the annual American Mathematical Society meetings and presents his research at international conferences such as the Conference on Mathematical Modeling and Analysis and the Volga Neuroscience Meeting. Recently Michael published four manuscripts on international pedagogy and has been invited as one of the keynote speakers at several international and interdisciplinary conferences such as the International Scientific Conference Society, Integration and Education, held annually at the Rezekne Technical Academy in Latvia. He taught courses and conducted seminars on these related topics during his spring 2009 sabbatical at the Aegean University in Greece and during his spring 2016 sabbatical at Riga Technical University in Latvia. In addition, Michael taught a new course on "Introduction to Recognition and Deciphering of Patterns" at the Rezekne Technical Academy in Rezekne, Latvia, in May 2019. Michael's aim is to inspire students to learn.

Recently, Michael had the opportunity to implement his hands-on teaching and learning style in the courses that he regularly teaches at RIT and during his spring 2016 sabbatical in Latvia. This method has proven to work successfully for him and his students, keeping the students stimulated and engaged and improving their course performance [23,26].

During his spare time Michael spends time outdoors and is an avid landscape photographer. In addition, he is an active poet and has several published poems about nature in the *LeMot Juste*. Furthermore, Michael published an article on "Re−Photographing the Baltic Sea Scenery in Liepaja: Why photograph the same scenery multiple times" in the *Journal of Humanities and Arts 2018* and a book on "Poetic Landscape Photography" with *JustFiction Edition 2019*. Spending time outdoors and active landscape photography widens and expands his understandings of nature's patterns and cadences.

CHAPTER 1

Introduction

W E DEFINE A **PATTERN** as "repeated design or recurring sequence, an ordered set of numbers, shapes or other mathematical objects arranged according to a rule." Recognition of patterns is a pertinent part of the learning process as we observe patterns during different occasions such as traffic patterns, musical patterns, language patterns, behavioral patterns in psychology, patterns in decision making, nature's formations patterns, weather patterns (winds, storms, waves, etc.), patterns in computer programs, signal patterns and signal processing, decorative patterns, architectural patterns and mathematical patterns. Sometimes patterns repeat at the same scale and, at other times, patterns repeat at different scales. We can also discover alternating patterns. We will begin with examples of nature's patterns (Figure 1.1). The first example portrays the repetitive alpine patterns of the Rocky Mountains in Rocky Mountain National Park:

FIGURE 1.1   The Rocky Mountains render repeated patterns at the same scale.

The next example is an aerial photograph that details a system of repeated canyons at different scales (Figure 1.2):

FIGURE 1.2   System of canyons depicting repeated patterns at different scales.

The consequent aerial photograph renders an alternating theme of the clouds and their reflections in the Atlantic Ocean (Figure 1.3):

FIGURE 1.3   Clouds and their reflections in the Atlantic Ocean.

The book's aim is to get acquainted with how and when various patterns arise and how to describe distinct patterns geometrically and analytically with guided repetitive-type examples. Throughout this book we will find assorted patterns in diverse geometrical applications and will apply distinct patterns as an essential counting technique. In some instances, we will discover that it will not be possible to acquire a formula for a specific pattern in one fragment and will require to decompose into piecewise patterns (a pattern that consists of at least two different sub-patterns or sub-systems). In addition, we will decompose a specific pattern into sub-patterns or into sub-systems of patterns.

The intent of this book is to provide several guided repetitive-type examples as it will enhance our understanding of patterns' fundamentals and recursive sequences; how and when they emerge; their distinct patterns; and their applications. In fact, several guided repetitive-type examples will help us develop our intuition on pattern recognition and help us see the larger spectrum on how concepts relate to each other that will lead to the development of theorems and their proofs. This is an essential technique that will direct us to the proof by induction method and will be used to generalize numerous results.

1.1 PATTERNS IN GEOMETRICAL SYSTEMS

We will commence with our journey on patterns' emergence with several geometry examples and provide varieties of examples that characterize decomposition of figures at the same scale and **geometrical fractals** that resemble decomposition of figures at diminishing scales. The upcoming example renders a system of identical equilateral triangles replicated at the same scale.

Example 1.1. *In this example we decompose the primary equilateral triangle into smaller equilateral triangles at the same scale as illustrated in corresponding diagram:*

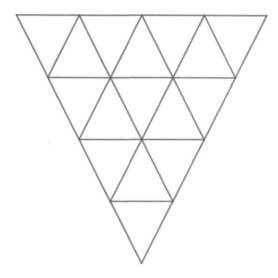

FIGURE 1.4   System of decomposed equilateral sub-triangles.

Notice that in Figure (1.4) there are four rows of triangles and each row has an odd number of triangles. In fact, the very bottom row consists of one triangle, the next row above has three triangles, the consequent row above has five triangles and so on. In Chapter 2, we will remit the following vital question: How many sub-triangles are inside the primary triangle? This will require the knowledge of a specific pattern whose sum adds up to a perfect square; we will examine this pattern more thoroughly in Chapters 2 and 3. In Chapter 2, we will apply piecewise functions to design and to describe the assigned system of triangles in Figure (1.4).

Similar to Example (1.1), the upcoming example will transition to a slightly different version of a system of equilateral triangles with alternating green and blue colors and with alternating triangular shapes.

Example 1.2. *In this example, we will break up the principal equilateral triangle into smaller equilateral triangles at the same scale with alternating green and blue colors and with alternating triangular shapes as evinced in the diagram below:*

FIGURE 1.5   System of alternating green and blue equilateral triangles.

Figure (1.5) portrays a system of equilateral triangles painted with blue and green colors. First of all, any two neighboring triangles in the horizontal and vertical directions have different colors. Second of all, any two neighboring triangles emerge in different directions; the green triangles face downward and the blue triangles face upward. Therefore, the green and blue triangles are sub-patterns of the primary triangle.

In Chapters 2 and 3, we will address a similar question as found in Example (1.2): How many green triangles and how many blue triangles are inside the principal triangle? Recognition of a certain pattern that sums the consecutive positive integers will determine the exact number of triangles. Furthermore, we can consider this an alternating pattern due to fetching of the colors from green to blue and due to the alternating triangular shapes.

Now we will shift gears to geometrical examples of Fractals. We define a **Fractal** as a repetition of patterns at different scales. The upcoming example will depict a system of diminishing triangles.

Example 1.3. *In this example, the dominant blue equilateral triangle is decomposed into smaller equilateral triangles by connecting the lines between the three midpoints of each triangle as illustrated below:*

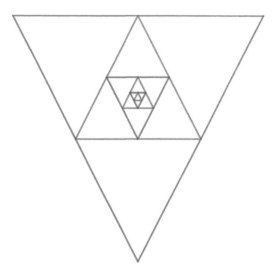

FIGURE 1.6   System of diminishing equilateral triangles.

In Figure (1.6) we construct a specific system of shrinking triangles by connecting the lines of the green triangle between the midpoints of the main blue triangle and so on. Observe that the green triangles and blue triangles decrease in opposite patterns. In fact, the green triangle is assembled by reflecting the blue triangle vertically while it declines in size. How many triangles are inscribed inside the blue focal triangle? This is rendered by a specific **linear sequence***, or a* **linear pattern***.*

The dimensions of the triangles, on the other hand, are described by a **geometric sequence***, or a geometrical pattern. This is also an example of an* **alternating sequence***, or an alternating pattern as the shapes of the triangles alternate. In Chapter 2, we will apply horizontal, vertical and diagonal lines on restricted intervals to describe the assigned system of diminishing triangles in Figure (1.6).*

The next fractal example portrays a system of diminishing circles.

Example 1.4. *We commence with inscribing four green circles inside the dominant blue circle as demonstrated below:*

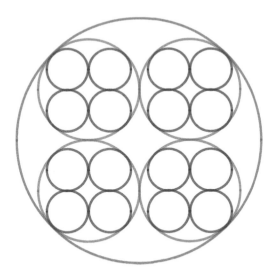

FIGURE 1.7   System of diminishing circles.

Notice that in Figure (1.7) the principal blue circle contains four green circles and each green circle has four red circles and so on. If we were to proceed with this algorithm by inscribing four smaller circles inside the dominant circle, then how many total circles are there? We will address this question in Chapter 3 and will use a specific **geometric summation** *to determine the total number of circles.*

We will examine more detailed examples of related patterns to geometrical systems in Chapters 2 and 3 and how to describe these patterns analytically with linear patterns and geometric patterns.

1.2 PIECEWISE FUNCTIONS AND PATTERNS

We will advance with our inquiry of patterns that appear as piecewise functions graphically. We define a **Piecewise Function** as a function that consists of two or more functions (fragments) on different restricted intervals. Piecewise functions have numerous applications in describing economic cycles, population dynamics, weather patterns, signal processing, neural networking and other natural phenomena. We will examine assorted examples of piecewise functions that will guide us to various recognition of patterns.

Example 1.5. *The following graph describes the positive integers as a system of horizontal lines on restricted intervals with length 1 starting with the horizontal line $y = 1$ on the interval $(0, 1)$, $y = 2$ on the interval $(1, 2)$ and so on as illustrated in the corresponding sketch:*

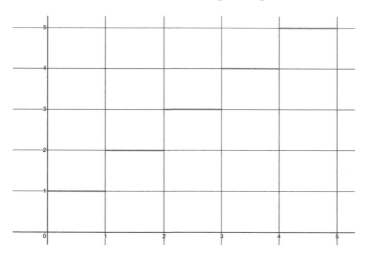

FIGURE 1.8   System of escalating horizontal lines.

Notice that in Figure (1.8) the steps escalate in increments of one from one neighboring interval to the other. Therefore, for all $n \in \mathbb{N}$ we can write the corresponding piecewise equation that depicts Figure (1.8):

$$
y = \begin{cases}
1 & \text{if } x \in (0, 1) \\
2 & \text{if } x \in (1, 2) \\
3 & \text{if } x \in (2, 3) \\
4 & \text{if } x \in (3, 4) \\
\vdots \\
n & \text{if } x \in (n - 1, n) \\
\vdots
\end{cases}
$$

Furthermore, Figure (1.8) and its corresponding piecewise equation list all the consecutive positive integers (natural numbers $\mathbb{N}$) starting with 1:

$$1, \ 2, \ 3, \ 4, \ 5, \ 6, \ldots .$$

*This **linear pattern** is rendered in Example (1.6). We also encountered this pattern in Figure (1.5) in Example (1.2). We will discuss deeper specifics of this pattern and related patterns in the next section and in later chapters.*

Similar piecewise graphs can be assembled with their corresponding patterns and will be left as end-of-chapter exercises in Chapter 2. The next example renders an oscillatory and alternating character between the horizontal lines $y = 1$ and $y = -1$ on intervals with length 1.

Example 1.6. *The ensuing graph portrays a piecewise function oscillating and alternating between the horizontal lines $y = 1$ and $y = -1$ on intervals with length 1:*

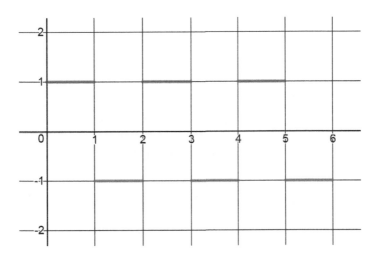

FIGURE 1.9   System of alternating horizontal lines.

*Note that Figure (1.9) commences with $y = 1$ on the interval $(0, 1)$, switches to $y = -1$ on the interval $(1, 2)$ and so on and renders the corresponding **alternating pattern**:*

$$1, \ -1, \ 1, \ -1, \ 1, \ -1, \ \dots . \tag{1.1}$$

*Hence for all $n \geq 0$, we express Eq. (1.1) and Figure (1.9) as the related **piecewise function**:*

$$
y = \begin{cases}
1 & \text{if } x \in (0,1) \\
-1 & \text{if } x \in (1,2) \\
1 & \text{if } x \in (2,3) \\
-1 & \text{if } x \in (3,4) \\
\ \vdots \\
1 & \text{if } x \in (2n, 2n+1) \\
-1 & \text{if } x \in (2n+1, 2n+2) \\
\ \vdots
\end{cases}
$$

For all $n \geq 0$, we can alternatively rewrite the above expression as:

$$
y = (-1)^n \text{ if } x \in (n, n+1).
$$

Analogous piecewise functions can be contrived and will be left as end-of-chapter exercises. We will proceed with a slightly different example of a piecewise function that renders diagonal lines with positive slope 1 and with negative slope -1 together with its' characteristics. This is also an example of an alternating pattern or sequence. Additional details of alternating sequences will be studied in Chapters 3 and 6.

Example 1.7. *The next graph depicts an oscillatory trait of a system of blue diagonal lines with positive slope 1 and green diagonal lines with negative slope -1:*

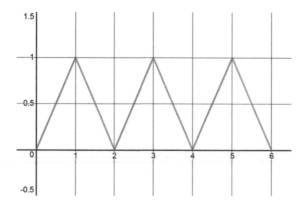

FIGURE 1.10 System of diagonal lines with alternating positive and negative slopes.

Notice that Figure (1.10) starts with $y = x$ on the interval $[0, 1]$ and switches to $y = 2 - x$ on the interval $[1, 2]$ and so on. Observe that the blue parallel lines have a positive slope 1 and the green parallel lines have a negative slope -1. Hence for all $n \geq 0$ we acquire:

$$
y = \begin{cases}
x & \text{if } x \in [0, 1] \\
2 - x & \text{if } x \in [1, 2] \\
x + 2 & \text{if } x \in [2, 3] \\
4 - x & \text{if } x \in [3, 4] \\
x + 4 & \text{if } x \in [4, 5] \\
\vdots & \\
x + 2n & \text{if } x \in [2n, 2n + 1] \\
(2n + 2) - x & \text{if } x \in [2n + 1, 2n + 2] \\
\vdots &
\end{cases}
$$

Notice that the y-intercepts of the blue and green lines rise in increments of 2 and list all consecutive positive even integers that are depicted in Example (1.7). This is also an example of an **alternating pattern** *as two neighboring lines have slopes of opposite signs. Comparable piecewise functions can be devised and will be end-of-chapter exercises in Chapter 2.*

Now we will transition to describing patterns (sequences) analytically as a list of specific values that we analogously encountered in Examples (1.5) and (1.6).

1.3 DESCRIBING PATTERNS ANALYTICALLY

We will progress with our exploration of patterns by describing them analytically with a formula or with a set of formulas that generates a precise list of values (sequence). We will commence with introducing several definitions.

Definition 1. *For $n \in \mathbb{N}$, we define a* **Finite Sequence** $\{x_i\}_{i=1}^{n}$ *as a finite list of n values as:*

$$\{x_i\}_{i=1}^{n} = x_1, \ x_2, \ x_3, \ x_4, \ \ldots, \ x_{n-1}, \ x_n, \tag{1.2}$$

where 1 is the **starting index** *and n is the* **terminating index.** *We can rewrite an equivalent formula of Eq. (1.2) by shifting the starting and terminating indices by 1 less as:*

$$\{x_i\}_{i=0}^{n-1} = x_0, \ x_1, \ x_2, \ x_3, \ \ldots, \ x_{n-2}, \ x_{n-1}.$$

Similarly, by shifting the starting and terminating indices by 1 more, we rewrite Eq. (1.2) as:

$$\{x_i\}_{i=2}^{n+1} = x_2, \ x_3, \ x_4, \ x_5, \ \ldots, \ x_n, \ x_{n+1}.$$

Definition 2. *We define an* **Infinite Sequence** $\{x_n\}_{n=1}^{\infty}$ *as a nonterminating list of values in the form:*

$$\{x_n\}_{n=1}^{\infty} = x_1, \ x_2, \ x_3, \ x_4, \ \ldots, \ x_n, \ \ldots. \tag{1.3}$$

Definition 3. *For all $n \geq 0$, we define a* **Piecewise Sequence** $\{x_n\}_{n=0}^{\infty}$ *that consists of two sub-sequences $\{a_n\}_{n=0}^{\infty}$ and $\{b_n\}_{n=1}^{\infty}$ as:*

$$a_0, \ b_1, \ a_2, \ b_3, \ \ldots. \tag{1.4}$$

Notice that Eq. (1.4) can be expressed as:

$$\{x_n\}_{n=0}^{\infty} = \begin{cases} a_n & \text{if } n \text{ vis even,} \\ b_n & \text{if } n \text{ is odd.} \end{cases} \tag{1.5}$$

We will examine contrasting patterns such as Linear Sequences, Geometric Sequences, Summation-Type Sequences, Factorial, Alternating Sequence and Piecewise Sequences later in this chapter and in meticulous details in Chapters 2 and 3. The upcoming example will depict a **Linear Sequence** or a Linear Pattern.

Example 1.8. *In Example (1.5) we obtained the corresponding pattern that lists all the consecutive positive integers (natural numbers $\mathbb{N}$) starting at 1:*

$$\{n\}_{n=1}^{\infty} = 1, \ 2, \ 3, \ 4, \ 5, \ 6, \ 7, \ 8, \ \ldots. \tag{1.6}$$

Observe that Eq. (1.6) is assembled by adding a 1 from neighbor to neighbor. The difference between two neighbors is always 1. This is a special case of a **Linear Pattern** *when the difference between two neighbors is* *always a constant. We can then conceive numerous analogous Linear Patterns such as the consecutive positive even integers or the positive multiples of 2:*

$$\{2n\}_{n=1}^{\infty} = 2, 4, 6, 8, 10, 12, 14, 16, \ldots \quad (1.7)$$

Note that the difference between two neighbors is always 2 and Eq. (1.7) describes the consecutive positive even integers in blue below as a subsequence of:

$$1, 2, 3, 4, 5, 6, 7, 8, 9, 10, \ldots$$

Similar to Eq. (1.7), the next sequence lists the consecutive positive odd integers:

$$\{2n-1\}_{n=1}^{\infty} = 1, 3, 5, 7, 9, 11, 13, 15, \ldots \quad (1.8)$$

Similar to Eqs. (1.6)–(1.8), we can formulate analogous linear patterns formulated and will be end-of-chapter exercises and will be studied in Chapter 3.

The subsequent definition will define a **Geometric Sequence**.

Definition 4. *For $n \geq 0$, we define a* **Finite Geometric Sequence** *consisting of $n+1$ values as:*

$$\{a \ r^i\}_{i=0}^{n} = a, a \cdot r, a \cdot r^2, a \cdot r^3, a \cdot r^4, a \cdot r^5, \ldots, a \cdot r^n, \quad (1.9)$$

where a is the starting term of the sequence in Eq. (1.9) and r is the multiplicative factor. Notice that the pattern in Eq. (1.9) is erected by multiplying by r from neighbor to neighbor. The quotient of two neighbors is always r.

Example 1.9. *Similar to Example (1.8), the geometric sequence:*

$$\{4^n\}_{n=0}^{\infty} = 1, 4, 16, 64, 256, 1024, \ldots,$$

is a sub-sequence of:

$$\{2^n\}_{n=0}^{\infty} = 1, 2, 4, 8, 16, 32, 64, \ldots,$$

emphasized in blue, which is also a geometric sequence.

Examples (1.3) and (1.4) resemble the Geometric Sequence. For instance, in Example (1.3), the green triangle's dimensions are half of the principal blue triangle's dimensions and so on. In Example (1.4), the radius of the primary blue circle is four times larger than the green circles'; each green circle's radius is four times larger than the red circles' and so on. The next example exhibits an application of a geometric sequence in paper folding.

Example 1.10. *The following diagram evokes a square with area 1 folded in half (vertically first, then horizontally, etc.) eight times (Figure 1.11):*

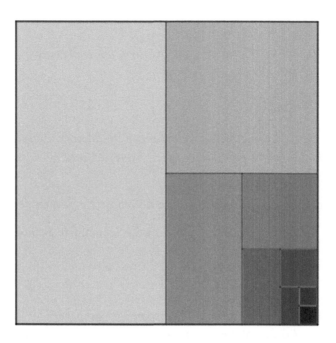

FIGURE 1.11   A square folded in half vertically first, then horizontally, etc.

First we cut the main square in half with a vertical red line, then with a horizontal red line, then with a red vertical line, then with a red horizontal line, etc. Diminishing rectangles and their associated areas are rendered with the darker shades of blue. During each fold we reduce the area by half and generate the following geometric sequence.

$$1, \frac{1}{2}, \left(\frac{1}{2}\right)^2, \left(\frac{1}{2}\right)^3, \left(\frac{1}{2}\right)^4, \left(\frac{1}{2}\right)^5, \left(\frac{1}{2}\right)^6, \left(\frac{1}{2}\right)^7, \left(\frac{1}{2}\right)^8$$

$$= \left\{ \left(\frac{1}{2}\right)^i \right\}_{i=0}^{8}.$$

This is another example of a Geometrical Fractal. In Chapter 2 we will describe a system of horizontal and vertical lines on corresponding restricted intervals that produces this fractal shape in Figure (1.11).

We will also assemble sequences as a **Sum** and as a **Product**.

1.3.1 Summation-Type and Product-Type Sequences

Definition 5. *For $n \in \mathbb{N}$, we define the **Sum** consisting of n values with the symbol $\sum$ as:*

$$S = x_1 + x_2 + x_3 + \ldots + x_n = \sum_{i=1}^{n} x_i.$$

The consequent example describes a sequence that adds the consecutive positive integers starting with 1 (natural numbers $\mathbb{N}$).

Example 1.11. *The corresponding **Summation-Type** pattern evokes the sum of natural numbers $\mathbb{N}$ starting with 1:*

$$x_1 = 1,$$
$$x_2 = 1 + 2,$$
$$x_3 = 1 + 2 + 3,$$
$$x_4 = 1 + 2 + 3 + 4,$$
$$x_5 = 1 + 2 + 3 + 4 + 5,$$
$$\vdots$$
$$x_n = 1 + 2 + 3 + 4 + \ldots + (n - 1) + n = \sum_{i=1}^{n} i.$$

Chapter 3 will render supplemental examples of **Summation-Type** sequences. The upcoming example will list the **Fibonacci Numbers** or the **Fibonacci Sequence**.

Example 1.12. *The subsequent pattern describes the **Fibonacci Numbers**:*

$$1, \ 1, \ 2, \ 3, \ 5, \ 8, \ 13, \ 21, \ \ldots . \tag{1.10}$$

The Fibonacci Sequence in Eq. (1.10) alternatively can be composed as a recursive sequence and as a summation-type sequence:

$$x_0 = 1,$$
$$x_1 = 1,$$
$$x_2 = x_0 + x_1 = 1 + 1 = 2,$$
$$x_3 = x_1 + x_2 = 1 + 2 = 3,$$
$$x_4 = x_2 + x_3 = 2 + 3 = 5,$$
$$x_5 = x_3 + x_4 = 3 + 5 = 8,$$
$$\vdots$$

The **Fibonacci Sequence** emerges in nature's numerous patterns such as the structure of pine cones, trees, leaves, sunflowers, seashells, waves, storm systems and the galaxies' formations. They often appear in the corresponding spiraling shape (Figure 1.12):

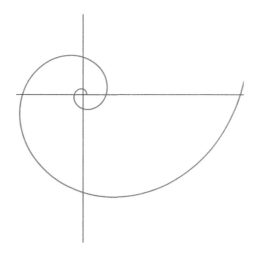

FIGURE 1.12   Fibonacci numbers in a spiral shape.

For instance, the consequent photograph renders the Fibonacci Pattern of the trees' structure starting with the smallest branches (Figure 1.13):

FIGURE 1.13   The Fibonacci sequence in nature.

Definition 6. *For* $n \in \mathbb{N}$, *we define the* **Product** *of* n *values with the symbol* $\prod$ *as:*

$$P = x_1 \cdot x_2 \cdot x_3 \cdot \ldots \cdot x_n = \prod_{i=1}^{n} x_i.$$

Example 1.13. *The* **Factorial Pattern** *is defined as a product of the consecutive positive integers starting with 1 as:*

$$0! = 1,$$
$$1! = 1,$$
$$2! = 2 \cdot 1 = 2 \cdot 1!,$$
$$3! = 3 \cdot 2 \cdot 1 = 3 \cdot 2!,$$
$$4! = 4 \cdot 3 \cdot 2 \cdot 1 = 4 \cdot 3!,$$
$$\vdots$$
$$n! = n \cdot (n-1) \cdot (n-2) \cdot \ldots \cdot 2 \cdot 1 = n \cdot (n-1)! = \prod_{i=1}^{n} i.$$

The Geometric Sequence becomes a special case of the Factorial when we multiply by the same constant r. The Factorial will be described as a recursive sequence in Chapter 5, and the Factorial emerges in **Pascal's Triangle** (Figure 1.14):

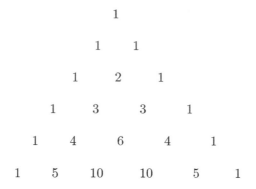

FIGURE 1.14   The Pascal's triangle with six rows.

In Chapter 4 we will examine deeper features of **Pascal's Triangle** by deciphering the triangle's rows and diagonals. We will apply identities expressed in combinations in proving specific patterns of the **triangle**.

Example 1.14. *For all $n \in \mathbb{N}$, rewrite the sequence $\{(-1)^{n+1}n\}_{n=1}^{\infty}$ as a piecewise sequence.*

Solution: *We group the terms of the given sequence into even-indexed terms in blue and into odd-indexed terms in green as:*

$$\{(-1)^{n+1}n\}_{n=1}^{\infty} = 1, -2, 3, -4, 5, -6, \ldots. \qquad (1.11)$$

For all $n \geq 0$, we can rewrite Eq. (1.11) as:

$$\{x_n\}_{n=0}^{\infty} = \begin{cases} [n+1] & \text{if } n \text{ is even,} \\ -[n+1] & \text{if } n \text{ is odd.} \end{cases}$$

We will seek and study additional examples of piecewise sequences and their applications in Chapters 2 and 3.

1.3.2 Piecewise and Periodic Patterns

Now we will shift our analysis of piecewise or alternating patterns (sequences) that consist of two or more fragments as we encountered in Examples (1.6) and (1.7) and that render periodic behavior and patterns.

Definition 7. *The sequence $\{x_n\}_{n=0}^{\infty}$ is periodic with **period-p**, $(p \geq 2)$, provided that for all $n \geq 0$:*

$$x_{n+p} = x_n.$$

The next two examples will portray piecewise, alternating and periodic patterns (sequences).

Example 1.15. *From Example (1.6) we acquired the following alternating or piecewise period-2 pattern:*

$$1, \; -1, \; 1, \; -1, \; 1, \; -1, \; \ldots. \tag{1.12}$$

For all $n \geq 0$, we can assert Eq. (1.12) with the following piecewise formula:

$$(-1)^n = \begin{cases} 1 & \text{if } n \text{ is even,} \\ -1 & \text{if } n \text{ is odd.} \end{cases}$$

The associated graph depicts the assigned period-2 cycle in (1.12) (Figure 1.15):

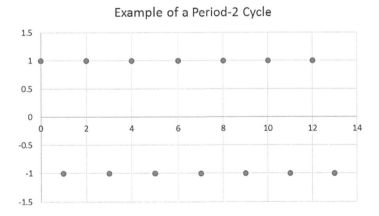

FIGURE 1.15 Alternating period-2 cycle oscillating between 1 and -1.

The next two examples will exhibit supplementary periodic traits with period-3 and period-4 and how piecewise sequences naturally emerge. This will guide us to investigate additional characteristics of piecewise sequences in Chapter 3. In addition we will discover deeper understanding of periodic traits of linear recursive sequences in Chapter 6 that depict various periods and their distinct patterns.

Example 1.16. *The corresponding graph below (Figure 1.16):*

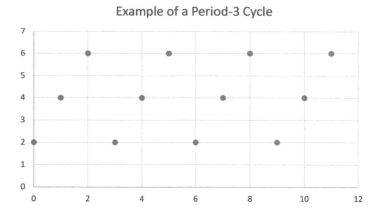

FIGURE 1.16   An increasing period-3 cycle with terms 2, 4 and 6.

renders the subsequent increasing period-3 cycle:

$$2,\ 4,\ 6,\ 2,\ 4,\ 6,\ \ldots,$$

which can also be expressed as the following piecewise sequence:

$$\{x_n\}_{n=0}^{\infty} = \begin{cases} 2 & \textit{if } n=3k, \\ 4 & \textit{if } n=3k+1, \\ 6 & \textit{if } n=3k+2. \end{cases}$$

Figure (1.16) renders a special case of periodic attributes of Eq. (6.10) in Chapter 6 when $x_0 = 1$, $b_0 = 2$, $b_1 = 2$ and $b_2 = -4$. In this specific instance we obtain a linear pattern that describes the consecutive even integers. On the other hand, this may not occur if we choose different values of b_0, b_1 and b_2. We will seek additional examples of periodicity in Chapter 6 and will compare the similarities and differences with periodic features in Figure (1.16).

Example 1.17. *The graph below (Figure 1.17):*

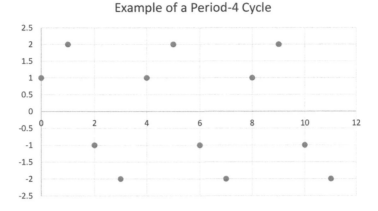

FIGURE 1.17   Period-4 cycle alternating between 1, 2, −1 and −2.

describes the corresponding alternating period-4 cycle:

$$1, \ 2, \ -1, \ -2, \ 1, \ 2, \ -1, \ -2, \ \ldots,$$

which can also be expressed as the analogous piecewise pattern:

$$\{x_n\}_{n=0}^{\infty} = \begin{cases} 1 & \text{if } n = 4k, \\ 2 & \text{if } n = 4k + 1, \\ -1 & \text{if } n = 4k + 2, \\ -2 & \text{if } n = 4k + 3. \end{cases}$$

Figure (1.17) is a special case of Eq. (6.6) in Chapter 6 when $x_0 = 1$, $a_0 = 2$ and $a_1 = -0.5$. We will examine supplemental examples of analogous periodic traits in Chapter 6 that mimic the pattern we see in Figure (1.17).

Examples (1.15) and (1.17) render alternating patterns and guide us to the following definition.

Definition 8. *The sequence $\{x_n\}_{n=0}^{\infty}$ is an **alternating periodic sequence** if for some $k \in \mathbb{N}$, $x_n = -x_{n+k}$ for all $n \geq 0$.*

Figure (1.15) depicts an alternating period-2 cycle as $x_{n+1} = -x_n$ for all $n \geq 0$. In addition, observe that Figure (1.17) depicts an alternating period-4 cycle as $x_{n+2} = -x_n$ for all $n \geq 0$. On the other hand, Figure (1.16) does not render an alternating period-3 cycle as all the terms of the cycle are positive. This will guide us to study further characteristics of alternating-periodic patterns characterized by recursive sequences in Chapter 6.

1.4 CHAPTER 1 EXERCISES

In problems 1–10, write a **formula** of each linear sequence:

  1:   3, 6, 9, 12, 15, 18, 21,....

  2:   7, 14, 21, 28, 35, 42, 49,....

  3:   15, 30, 45, 60, 75, 90, 105,....

  4:   8, 10, 12, 14, 16, 18, 20,....

  5:   20, 24, 28, 32, 36, 40, 44,....

  6:   5, 7, 9, 11, 13, 15, 17,....

  7:   4, 7, 10, 13, 16, 19, 22,....

  8:   2, 11, 20, 29, 38, 47, 56,....

  9:   4, 11, 18, 25, 32, 39, 46,....

 10: 1, 5, 9, 13, 17, 21, 25,....

In problems 11−18, write a **formula** of each geometric sequence:

 11: 8, 16, 32, 64, 128, 256, 512,....

 12: 6, 12, 24, 48, 96, 192, 384,....

 13: $\dfrac{1}{9}, \dfrac{1}{3}$, 1, 3, 9, 27, 81,....

 14: $\dfrac{3}{8}, \dfrac{3}{2}$, 6, 24, 96, 384, 1536,....

 15: 5, 15, 45, 135, 405, 1215, 3645,....

 16: 32, 8, 2, $\dfrac{1}{2}, \dfrac{1}{8}, \dfrac{1}{32}$, ...

 17: 64, 48, 36, 27, $\dfrac{81}{4}, \dfrac{343}{16}$, ...

 18: 9, 6, 4, $\dfrac{8}{3}, \dfrac{16}{9}, \dfrac{32}{27}$, ...

In problems 19−26, write a **formula** of each piecewise sequence:

19: 4, −2, 4, −2, 4, −2, ...

20: −1, 2, 1, −2, −1, 2, 1, −2, ...

21: 1, 3, −1, 1, 3, −1, 1, 3, −1, ...

22: 2, 4, −1, −3, 2, 4, −1, −3, ...

23: 1, −2, 3, −4, 5, −6, 7, −8, ...

24: −2, 4, −6, 8, −10, 12, −14, 16, ...

25: 1, −2, 4, −8, 16, −32, 64, ...

26: 2, 4, 6, −8, 10, 12, 14, −16, ...

In problems 27−30:

27: Write a **formula** of the piecewise function below:

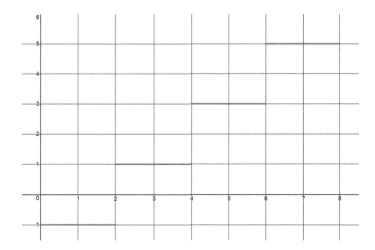

28: Write a **formula** of the piecewise function below:

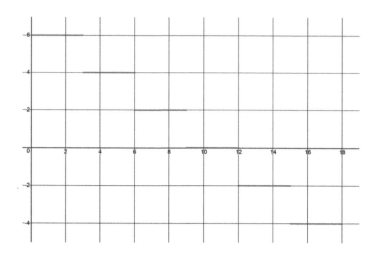

29: Write a **formula** of the piecewise function below:

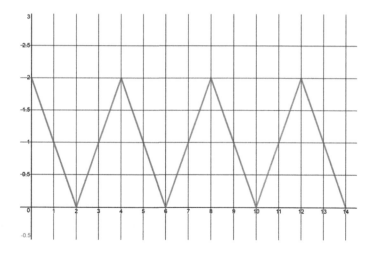

30: Write a **formula** of the piecewise function below:

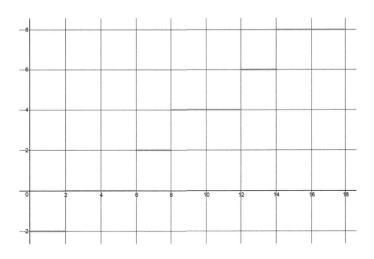

In problems 31−32:

31: Write a system of functions on restricted intervals that describe the right triangle below:

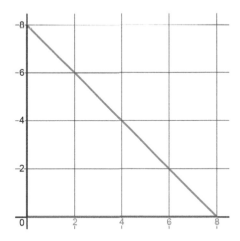

32: Write a system of functions on restricted intervals that describe the equilateral triangle below:

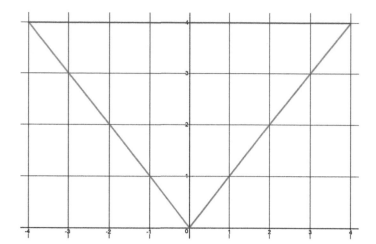

CHAPTER 2

Patterns of Geometrical Systems

THROUGHOUT THIS CHAPTER WE will render various examples describing how particular patterns (sequences) ensue from the assorted geometrical arrangements that repeat at the same scale, replicate at different scales (fractals) and alternating patterns of geometrical systems. During our exploration, we will address the following questions: How do we distinguish the similarities and differences between various geometrical arrangements? Why and how do certain arrangements guide us to specific patterns? In addition, we will examine the occurrence of patterns in **piecewise functions**. Analogous to Examples (1.1)–(1.4), we will examine distinct patterns (sequences) that originate from various geometrical arrangements and from contrasting piecewise functions.

Now we will begin with three diagrams that render various geometrical configurations and pose three related questions that ignite our curiosity on extraction about sequences. For instance, what patterns can emanate from the succeeding arrangement of right triangles as portrayed in the upcoming diagram? (Figure 2.1):

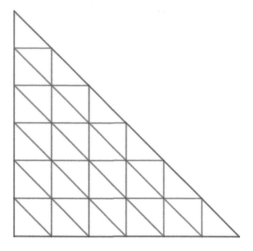

FIGURE 2.1   System of right triangles at the same scale.

Notice that this is the same pattern that we encountered in Figure (1.4) (adding consecutive positive odd integers). The next diagram remits the distinct patterns that arise by inscribing a circle inside a square and then a square inside a circle, etc., as portrayed in the corresponding sketch (Figure 2.2):

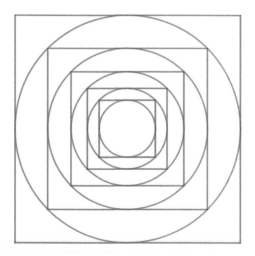

FIGURE 2.2   System of diminishing squares and circles.

This phenomenon is similar to Figure (1.6) as it is expressed by a geometrical pattern. What distinct pattern can we detect in the corresponding fractal system of diminishing triangles?

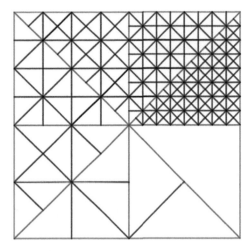

FIGURE 2.3 System of diminishing 45−45−90 triangles.

Note that the number of triangles increases from one neighboring region to the next between the red lines starting the largest triangle in the lower right corner. This occurrence is described by the geometric sequence. The sum of all the triangles is a **geometric sum** as portrayed in Figure (2.3).

2.1 REPETITION OF PATTERNS AT SAME SCALE

This section's objectives are to portray various examples of patterns that repeat at the same scale and trace their unique traits. We can observe repetition of patterns at the same scale in the architectural style of buildings. For instance, we can detect the repetitions of diverse patterns in Budapest's Parliament Building (Figure 2.4):

FIGURE 2.4   Hungary's Parliament Building in Budapest, Hungary.

The following photograph renders a replicated system of a bridges' arcs across the Venta River in Kuldiga, Latvia (Figure 2.5):

FIGURE 2.5   Bridge across the Venta River in Kuldiga, Latvia.

While spending time outdoors, we can discover the autumn's golden colors as the patterns reoccur at the same scale, depicted in the following photograph (Figure 2.6):

FIGURE 2.6   New Jersey's golden autumn scenery.

Furthermore, we can find the duplicated cascading patterns as shown in the corresponding photograph below (Figure 2.7):

FIGURE 2.7   Cascades in Franconia Notch State Park, New Hampshire.

Now we will progress with repetition of patterns at the same scale that emerge in piecewise functions.

2.2 PIECEWISE FUNCTIONS AT SAME SCALE

Examples (1.5)–(1.7) indicate repetition of patterns at same scale. We will examine more thorough examples of **piecewise functions** that portray replicated patterns at the same scale as combinations of horizontal lines and diagonal lines. We will work out the details in obtaining a formula in the first two examples. In the remaining four examples, writing a formula for each piecewise function will be left as an end-of-chapter exercise. Analogous to Example (1.5), the first example will render a sequence of positive odd integers as we encountered in Example (1.8) on the restricted intervals with length 2.

Example 2.1. *The following sketch depicts a sequence of positive odd integers on the restricted intervals with length 2 (Figure 2.8):*

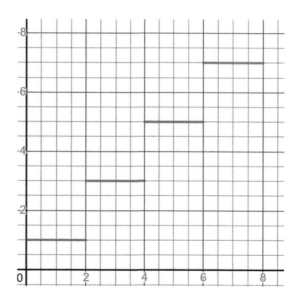

FIGURE 2.8 Sequence of positive odd integers on the restricted intervals with length 2.

Observe that Figure (2.9) commences with $y = 1$ on $(0, 2)$, $y = 3$ on $(2, 4)$, $y = 5$ on $(4, 6)$ and so on. Hence for all $n \geq 0$, we obtain the corresponding equation:

$$y = \begin{cases} 1 & \text{if } x \in (0, 2) \\ 3 & \text{if } x \in (2, 4) \\ 5 & \text{if } x \in (4, 6) \\ 7 & \text{if } x \in (6, 8) \\ \vdots & \\ 2n + 1 & \text{if } x \in (2n, 2n + 2) \\ \vdots & \end{cases}$$

Example 2.2. *The following sketch describes an alternating pattern between diagonal lines with a positive slope and horizontal lines:*

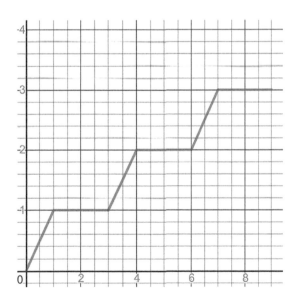

FIGURE 2.9 System of diagonal lines and horizontal lines.

Observe that the diagonal lines are on the restricted intervals with length 1 and the horizontal lines are on the restricted intervals with length 2. We start with $y = x$ on [0,1], $y = 1$ on [1,3], $y = x - 2$

on [3,4], $y = 2$ *on* [4,6] *and so on. Therefore, for all* $n \geq 0$ *we acquire the following formula:*

$$y = \begin{cases} x & \text{if } x \in [0, 1] \\ 1 & \text{if } x \in [1, 3] \\ x - 2 & \text{if } x \in [3, 4] \\ 2 & \text{if } x \in [4, 6] \\ \vdots & \\ x - 2n & \text{if } x \in [3n, 3n + 1] \\ n + 1 & \text{if } x \in [3n + 1, 3n + 3] \\ \vdots & \end{cases}$$

Example 2.3. *The consequent sketch stipulates a system of horizontal lines on the restricted intervals with length 1 and length 2 (Figure 2.10):*

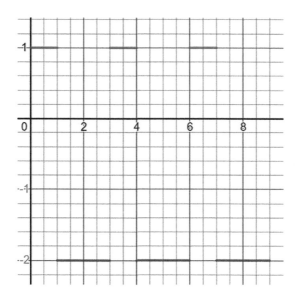

FIGURE 2.10 Piecewise function as a system of horizontal lines on the restricted intervals with length 1 and length 2.

Notice that the blue lines $y = 1$ *are on the restricted intervals with length 1 starting with the interval* $(0, 1)$*, while the red lines* $y = -2$ *are on restricted intervals with length 2 starting with the interval* $(1, 3)$*.*

Example 2.4. *The following sketch designates a system of horizontal lines and diagonal lines with a positive slope on the restricted intervals with length 1 (Figure 2.11):*

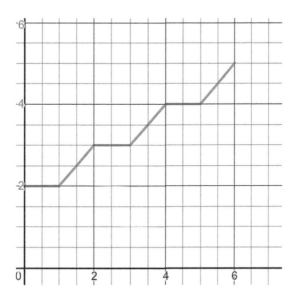

FIGURE 2.11 System of horizontal and diagonal lines on the restricted intervals with length 1.

The blue horizontal lines are on restricted intervals with length 1 starting with the interval $(0, 1)$, and the green diagonal lines (with slope 1) are on the restricted intervals with length 1 starting with the interval $(1, 2)$.

The next two examples extend from the fundamentals of Examples (2.3) and (2.4) with a specific system of horizontal and diagonal lines and various restricted intervals.

Example 2.5. *The corresponding graph depicts a system of horizontal lines and diagonal lines with a positive slope on the restricted intervals with length 1 and length 2 (Figure 2.12):*

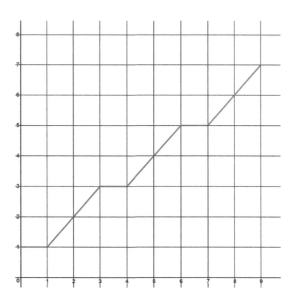

FIGURE 2.12 Horizontal and diagonal lines on the restricted intervals with length 1 and length 2.

The green horizontal lines are on the restricted intervals with length 1 starting with the interval $(0, 1)$, and the blue diagonal lines (with slope 1) are on the restricted intervals with length 2 starting with the interval $(1, 3)$.

Example 2.6. *The graph below traces a system of blue horizontal lines, green diagonal lines with a negative slope and red diagonal lines with a positive slope on the restricted intervals with length 1 and length 2 (Figure 2.13):*

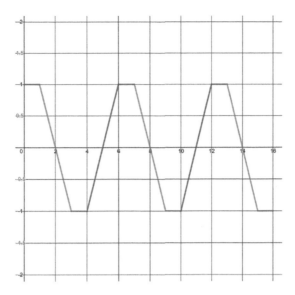

FIGURE 2.13 System of horizontal and diagonal lines on the restricted intervals with length 1 and length 2.

Analogous to Example (2.4), the blue horizontal lines are on the intervals with length 1 starting with the interval $(0, 1)$*, the green diagonal lines (with slope* -1*) are on the intervals with length 2 starting with the interval* $(1, 3)$ *and the red diagonal lines (with slope 1) are on the intervals with length 2 starting with the interval* $(4, 6)$*.*

Examples (2.3)–(2.6) describe the periodic traits with different periods. We will examine deeper periodic attributes of specific recursive sequences in Chapters 5 and 6. Now we will switch our focus to the geometrical examples that render the repetition of patterns at the same scale.

2.3 GEOMETRICAL CONFIGURATIONS AT SAME SCALE

Examples (1.1) and (1.2) in Chapter 1 render replicated shapes. We will proceed with further supplemental examples that resemble the reoccurring formations and the specific patterns. In addition, we will use piecewise functions to construct and describe certain systems of rectangles and triangles.

Example 2.7. *The following diagram depicts a square as a system of 36 squares assembled in either six rows, six columns or 11 diagonals:*

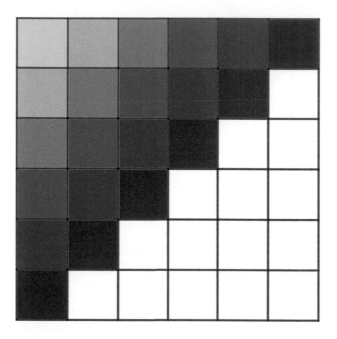

FIGURE 2.14   A 6 × 6 square resembled with six blue diagonals.

If we commence in Figure (2.14) from the light-blue square in the upper-left corner and include all the blue diagonals toward to the square's principal dark-blue diagonal, we obtain the pattern that enumerates all the consecutive positive integers between 1 and 6:

$$1,\ 2,\ 3,\ 4,\ 5,\ 6\ =\ \{i\}_{i=1}^{6}.$$

Furthermore, by adding all the blue diagonal squares we acquire the following sum:

$$1 + 2 + 3 + 4 + 5 + 6 = \sum_{i=1}^{6} i. \tag{2.1}$$

For all $n \in \mathbb{N}$, Eq. (2.1) extends to the following sum that adds all the consecutive positive integers:

$$1 + 2 + 3 + \ldots + (n-1) + n = \sum_{i=1}^{n} i = \frac{n[n+1]}{2}.$$

We will prove this summation formula using the **Proof By Induction** *technique in Chapter 3. Moreover, the corresponding system of horizontal lines and vertical lines on specific restricted intervals sketches the arrangement of squares in Figure (2.14):*

$$y = \begin{cases} 3 & \text{for } x \in [-3,3] \\ 2 & \text{for } x \in [-3,3] \\ \vdots \\ -2 & \text{for } x \in [-3,3] \\ -3 & \text{for } x \in [-3,3] \end{cases} \qquad x = \begin{cases} 3 & \text{for } y \in [-3,3] \\ 2 & \text{for } y \in [-3,3] \\ \vdots \\ -2 & \text{for } y \in [-3,3] \\ -3 & \text{for } y \in [-3,3] \end{cases}$$

Analogous to Example (1.1), the next example will exhibit a pattern of emulated right $45-45-90$ triangles.

Example 2.8. *The following sketch portrays the principal red right triangle structured as an arranged system of blue $45-45-90$ triangles:*

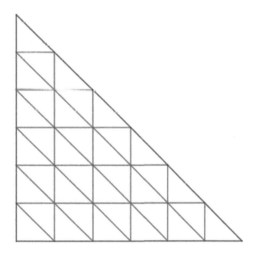

FIGURE 2.15   A system of replicated blue $45-45-90$ triangles.

First of all, analogous to the green and blue alternating triangles in Example (1.2), Figure (2.15) resembles an alternating pattern as two neighboring triangles are positioned in opposite directions (in horizontal and vertical directions). Second of all, the principal red right triangle is also a $45-45-90$ and consists of six rows.

Moreover, the first row has one triangle, the second row has three tri-angles, the third row has five triangles, the fourth row has seven triangles and so on. Therefore, each row has an odd number of triangles and traces the following pattern that describes the consecutive positive odd integers:

$$1, \ 3, \ 5, \ 7, \ 9, \ 11 \ = \ \{(2i - 1)\}_{i=1}^{6}.$$

Furthermore, by adding all the sub-right triangles we obtain the corresponding sum:

$$1 + 3 + 5 + 7 + 9 + 11 \ = \ \sum_{i=1}^{6} (2i - 1) \ = \ 36 \ = \ 6^2. \quad (2.2)$$

For all $n \in \mathbb{N}$, (2.2) extends to the following sum that adds all the consecutive positive odd integers:

$$1 + 3 + 5 + \ldots + (2n - 3) + (2n - 1) \ = \ \sum_{i=1}^{n} (2i - 1) \ = \ n^2.$$

*We will prove this summation formula using the **Proof By Induction** technique in Chapter 3. Sketching the decomposed arrangement of right triangles will require a system of horizontal lines, vertical lines and diagonal lines with a negative slope on various restricted intervals. The diagonal lines will be a family of parallel lines with the same slope but different y-intercepts. This will be left as an end of chapter exercise.*

The upcoming example is an extension of the base of Example (2.8) and will decompose the primary triangle as a system of alternating green and blue 45−45−90 right triangles. In comparison to Example (1.2) in Chapter 1, we will discover two cognate but distinct arising patterns as we did in Figure (1.5).

Example 2.9. *Analogous to Example (1.2) (alternating green and blue equilateral triangles in each row of the primary equilateral triangle), the primary 45−45−90 triangle is regrouped into rows of interchanging green and blue 45−45−90 triangles as illustrated in the corresponding diagram:*

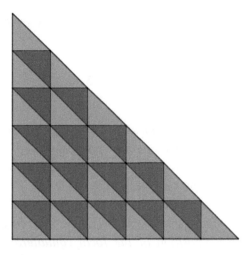

FIGURE 2.16   An alternating system of 45−45−90 green and blue triangles.

In Figure (2.16), the principal 45−45−90 triangle is contrived into six rows. In Example (2.8) we showed that every row contains an odd number of triangles. Also, in Figure (2.16) the green triangles appear in six rows and, hence, trace the pattern of the consecutive positive integers:

$$1, \ 2, \ 3, \ 4, \ 5, \ 6 \ = \ \{i\}_{i=1}^{6}.$$

Thus by adding all the green right triangles in Figure (2.16) we obtain the corresponding summation:

$$1 \ + \ 2 \ + \ 3 \ + \ 4 \ + \ 5 \ + \ 6 \ = \ \sum_{i=1}^{6} i.$$

The blue triangles, on the other hand, emerge in five rows and render the related consecutive positive integers' ordering:

$$1, \ 2, \ 3, \ 4, \ 5 \ = \ \{i\}_{i=1}^{5}.$$

Similarly by summing all the blue right triangles in Figure (2.16) we procure:

$$1 + 2 + 3 + 4 + 5 = \sum_{i=1}^{5} i.$$

The system of 45−45−90 triangles in Figure (2.16) is constructed by the combinations of parallel horizontal lines, parallel vertical lines and parallel diagonal lines with a negative slope on assorted restricted intervals. This will be left as end-of-chapter exercises.

We encountered these summations in Figure (2.14) in Example (2.7). In addition, we can detect exactly the same patterns in Figure (1.5) that characterize the decomposition of the focal equilateral triangle into an arrangement of alternating blue and green equilateral triangles.

The succeeding section will focus on **Fractals** or replicated patterns at different scales. We will perceive new patterns that govern the specific arrangements and compare the similarities and differences to what we experienced in this section.

2.4 REPETITION OF PATTERNS AT DIFFERENT SCALES

In Chapter 1 we defined a **Fractal** as the repetition of patterns at different scales and examined contrasting examples. This section's goals are to expand our knowledge of geometrical fractals as we encountered in Examples (1.3), (1.4) and (1.10) and render their unique traits and their associated patterns. We can observe the repetition of patterns at different scales while spending time outdoors: the coastal formations, the clouds' shapes, the alpine ridges and other fractal-type arrangements. For instance, while flying we can discern the Rockies' alpine patterns at diminishing scales as illustrated below (Figure 2.17):

FIGURE 2.17   An alpine system rendering decreasing sub-ridges.

Notice that the sub-ridges diminish as they split off from the main ridge and descend in elevation. At lower elevations we can see the formations at smaller scales and the bigger miens at higher elevations. The next marine photograph renders the system of waning clouds (Figure 2.18):

FIGURE 2.18   Clouds' miens above the Baltic Sea in Latvia.

It is interesting to note that the clouds' shapes gradually abate as they move further out into the sea. Sometimes the opposite phenomenon occurs when the smaller clouds are closer to the shore and the larger clouds are further out in the sea. Now we will advance with the deeper study of geometrical fractals and their distinct characteristics.

2.4.1 Geometrical Patterns at Different Scales

Example (1.3) describes the system of diminishing triangles starting with the primary triangle, and Example (1.4) depicts the system of diminishing circles emerging from the focal circle. In addition, Example (1.10) portrays paper folding and, hence, characterizes a **Finite Geometric Sequence** in the form:

$$\{a \cdot r^i\}_{i=0}^n = a, \, a \cdot r, \, a \cdot r^2, \, a \cdot r^3, \, \ldots, \, a \cdot r^n \,.$$

The chapter's and section's goals are to enhance our knowledge on the construction of geometrical fractals and their corresponding patterns. We will compare the similarities and differences with the patterns acquired in Chapter 1. The upcoming examples will evince the repetitions of patterns at diminishing scales described by a **geometric sequence**.

Example 2.10. *The following diagram resembles a system of diminishing squares:*

FIGURE 2.19　System of diminishing squares.

In Figure (2.19), we commence the algorithm with the primary green square and construct the neighboring blue square by connecting the diagonal blue lines between the mid-points of the primary green square. The consequent green square is designed by connecting the horizontal green lines between the mid-points of the blue square and so on. Now suppose that the length of each side of the primary green square is 1, then by applying the Pythagorean theorem and the properties of $45-45-90$ triangles, the length of each side of the neighboring blue square is $\frac{1}{\sqrt{2}}$. Hence for all $n \geq 0$, we acquire the related **geometric sequence** *that renders the assigned algorithm depicted in Figure (2.19):*

$$1, \frac{1}{\sqrt{2}}, \frac{1}{2}, \frac{1}{2\sqrt{2}}, \frac{1}{4}, \frac{1}{4\sqrt{2}}, \cdots, \left(\frac{1}{\sqrt{2}}\right)^n, \cdots \qquad (2.3)$$

Notice that the green terms of Eq. (2.3) render the length of each green square and the blue terms of the equation trace the length of each blue square. We can also rewrite Eq. (2.3) as a **piecewise sequence** *that consists of two geometric sequences.*

Furthermore, notice that the area of the primary green square is 1, and the area of the adjacent blue square is $\frac{1}{2}$. Hence for all $n \geq 0$ we procure the following **geometric sequence** *that describes the sequence of areas:*

$$1, \frac{1}{2}, \frac{1}{4}, \frac{1}{8}, \frac{1}{16}, \frac{1}{32}, \cdots, \left(\frac{1}{2}\right)^n.$$

Sketching Figure (2.19) will require the use of parallel horizontal, parallel vertical and parallel diagonal lines on the corresponding restricted intervals. In fact, the green squares are depicted by a system of parallel horizontal and vertical lines on the corresponding restricted intervals, while the blue squares are described by a system of parallel diagonal lines with a positive slope and negative slope on the corresponding restricted intervals. This will be left as end-of-chapter exercises.

The subsequent example will expand on Example (2.10) by determining the number of symmetrical triangles devised in Figure (2.19).

Example 2.11. *Our goal is to write the* **linear sequence** *that determines the total number of red triangles portrayed in the corresponding sketch:*

FIGURE 2.20   System of symmetrical red $45-45-90$ triangles.

Each time we incise a blue square inside a green square and vice versa we produce four symmetrical red $45-45-90$ triangles as shown in Figure (2.20). Therefore, for $n \in \mathbb{N}$, the total number of generated triangles mimics the corresponding **linear pattern***:*

$$4,\ 8,\ 12,\ 16,\ \dots,\ 4 \cdot n\ =\ \{4 \cdot i\}_{i=1}^{n}.$$

We will discover additional linear patterns that emerge in future examples that render similar geometrical configurations.

The next two examples will focus on a system of diminishing equilateral triangles whose dimensions and areas will be resembled by a **geometric sequence**.

Example 2.12. *From Example (1.3), the following diagram evokes a system of diminishing blue and green equilateral triangles:*

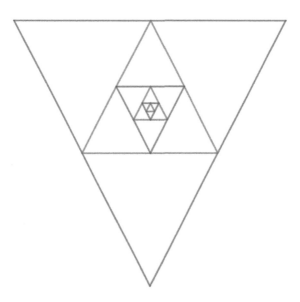

FIGURE 2.21   System of diminishing equilateral triangles.

In Figure (2.21) we assemble the first green triangle by connecting the green lines between the mid-points of the primary blue triangle and so on. Analogous to Example (2.10), the sides and the areas of the triangles will be characterized by a geometric sequence. Suppose that the length of each side the largest blue triangle is 2. Then the length of each side of the consequent green triangle is 1 as it is half the length of the blue triangle and so on. Therefore, for all $n \geq 0$ we acquire the following **geometric sequence** *that depicts the length of the sides of all the triangles starting with the primary blue triangle:*

$$2, \ 1, \ \frac{1}{2}, \ \frac{1}{2^2}, \ \frac{1}{2^3}, \ \frac{1}{2^4}, \ \cdots, \ \left(\frac{1}{2}\right)^{n-1}. \tag{2.4}$$

The blue terms of Eq. (2.4) render the side of each blue equilateral triangle, while the green terms of the equation trace the side of each green equilateral triangle. Now note that using the properties of $30-60-90$ triangles, the area of the primary blue triangle is $\sqrt{3}$, and the area of the adjacent green triangle is $\frac{\sqrt{3}}{4}$. Hence for all $n \geq 0$, we procure the following **geometric sequence** that describes the sequence of areas:

$$\sqrt{3}, \ \frac{\sqrt{3}}{4}, \ \frac{\sqrt{3}}{4^2}, \ \frac{\sqrt{3}}{4^3}, \ \frac{\sqrt{3}}{4^4}, \ \frac{\sqrt{3}}{4^5}, \ \ldots, \ \frac{\sqrt{3}}{4^n}.$$

Example 2.13. *Analogous to Example (2.11), our aim is to write a **linear sequence** that enumerates the total number of red triangles as illustrated in the corresponding sketch (Figure 2.22):*

FIGURE 2.22   System of red symmetrical equilateral triangles.

Note that we construct three red equilateral sub-triangles of the same size by linking the lines between the three mid-points of the principal blue triangle. Thus, for all $n \in \mathbb{N}$, the total number of red triangles including the smallest dark red triangle in the very center is rendered by following **linear sequence***:*

$$1, \ 4, \ 7, \ 10, \ \ldots, \ 3n + 1 \ = \ \{3i + 1\}_{i=0}^{n}.$$

The succeeding example will examine a system of diminishing squares and circles and will mimic similar dynamics in comparison to Example 2.10, where we inscribed a smaller square inside a bigger square. On the other hand, in the next example we will encase a circle inside a square instead.

Example 2.14. *The following diagram resembles a system of diminishing blue squares and green circles (Figure 2.23):*

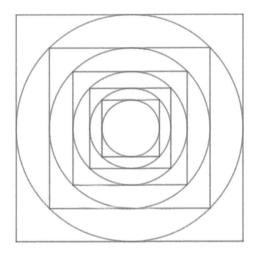

FIGURE 2.23   System of squares and circles.

We begin by inscribing a green circle inside the main blue square with side 2, whose corresponding area is 4. Notice that the circle's diameter is also 2 (radius 1). The diagonal of the next blue square inscribed the green circle is 2 and, hence, the length of the blue square is $\sqrt{2}$, whose corresponding area is 2. Analogous to Example 2.10, for all $n \geq 0$ the following **geometric sequence** *describes the length of the blue squares starting with the primary square:*

$$2, \sqrt{2}, 1, \frac{1}{\sqrt{2}}, \frac{1}{2}, \frac{1}{2\sqrt{2}}, \ldots, \left(\frac{1}{\sqrt{2}}\right)^{n-2}.$$

In addition, for all $n \geq 0$ we obtain the corresponding **geometric sequence** *that renders the area of the blue squares:*

$$4, 2, 1, \frac{1}{2}, \ldots, \left(\frac{1}{2}\right)^{n-2}.$$

Then for all $n \geq 0$, we produce the corresponding **geometric sequence** *that depicts the radii of the green circles commencing with the largest green circle:*

$$1, \frac{1}{\sqrt{2}}, \frac{1}{2}, \frac{1}{2\sqrt{2}}, \ldots, \left(\frac{1}{\sqrt{2}}\right)^n.$$

Finally for all for $n \geq 0$, we procure the **geometric sequence** *that describes the area of the green circles:*

$$\pi, \frac{\pi}{2}, \frac{\pi}{4}, \frac{\pi}{8}, \ldots, \frac{\pi}{2^n}.$$

In order to write a sequence describing the alternating pattern of the blue squares and the green circles at the same time will require a **piecewise geometric sequence** *that will consists of two geometric sequences. This concept will be addressed in thorough detail in Chapter 3.*

The upcoming example evokes a fractal system of a square with 45–45–90 triangles at diminishing scales. The goal is to determine the total number of 45–45–90 triangles inside the square. This will guide us to a geometric summation. Comparable phenomena occurred in Chapter 1 in Example (1.4) and in Example (1.10).

Example 2.15. *Analogous to Examples (1.4) and (1.10), the following diagram constructs an arrangement of a main blue square into a system of diminishing 45−45−90 triangles (Figure 2.24):*

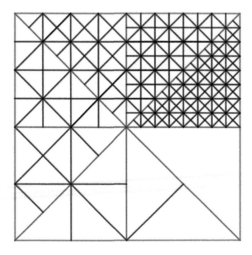

FIGURE 2.24   System of diminishing 45−45−90 triangles.

Starting from the largest 45−45−90 triangle, the exact number of 45−45−90 triangles that are inserted inside the blue square is characterized by the following **geometric sum***:*

$$
\begin{aligned}
& 1 + 2 + 4 + 8 + 16 + 32 + 64 + 128 \\
= \; & 1 + 2 + 2^2 + 2^3 + 2^4 + 2^5 + 2^6 + 2^7 \\
= \; & \sum_{i=0}^{7} 2^i = 2^8 - 1.
\end{aligned}
\tag{2.5}
$$

Observe Eq. (2.5) adds eight terms as the blue square is decomposed into eight primary triangular regions (emphasized by the red diagonal, horizontal and vertical lines) and into eight categories of 45−45−90 triangles. Hence, for all $n \in \mathbb{N}$, Eq. (2.5) extends to the corresponding **geometric summation***:*

$$
a + a \cdot r + a \cdot r^2 + a \cdot r^3 + \ldots + a \cdot r^n = \frac{a[1 - r^{n+1}]}{1 - r}.
\tag{2.6}
$$

Note that Eq. (2.6) has $n + 1$ terms added and $r \neq 1$. We will verify Eq. (2.6) by using the **Proof by Induction** *technique in Chapter 3.*

Example 2.16. *The following diagram renders the primary red 45-45-90 triangle decomposed into smaller 45-45-90 triangles:*

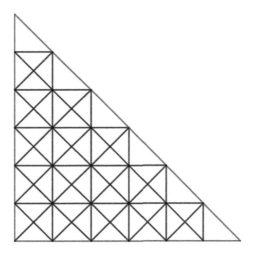

FIGURE 2.25   An arrangement of 45−45−90 triangles.

Notice that in Figure (2.25), the first row has one triangle, the second row has five triangles, the third row has nine triangles and so on. Therefore, each row has an odd number of triangles described by the following **linear sequence***:*

$$1, \ 5, \ 9, \ 13, \ 17, \ 21 \ = \ \{4i - 3\}_{i=1}^{6}.$$

The total number of triangles inscribed inside the principal red triangle is specified by the following sum:

$$1 + 5 + 9 + 13 + 17 + 21 \ = \ \sum_{i=1}^{6} (4i - 3). \qquad (2.7)$$

Observe that for all $n \in \mathbb{N}$, Eq. (2.7) extends to the following sum that adds the corresponding odd integers:

$$1 + 5 + 9 + 13 + \ldots + (4n - 3) \ = \ \sum_{i=1}^{n} (4i - 3) \ = \ n \cdot [2n - 1]. \ (2.8)$$

We will derive Eq. (2.8) and analogous formulas and prove them by applying the **Proof by Induction** *method in Chapter 3.*

2.5 ALTERNATING AND PIECEWISE PATTERNS

This section's aims are to render the occurrence of alternating patterns and their distinct attributes. We encountered several specimens of geometrical alternating patterns in Chapter 1 in Examples (1.2), (1.3), (1.6) and (1.7) and in this chapter in Examples (2.8), (2.10), (2.14) and (2.12). We will divulge into deeper details with deriving **piecewise sequences** that consist of two or more sequences to describe the specific phenomena in several examples from the previous section.

We will commence with examples of interchanging architectural patterns and nature's alternating alpine and marine patterns. For instance, the main building of Liepaja University in Liepaja, Latvia, renders red and white alternating colors and architectural styles and decorations as shown below (Figure 2.26):

FIGURE 2.26 Alternating architectural patterns of Liepaja's University main building.

During the holiday season, we can discern varying blue and white decorative colors as portrayed in the photograph below (Figure 2.27):

FIGURE 2.27 Alternating blue and white decorative lights in Riga, Latvia.

We can also detect nature's alternating schemes while traveling and spending time outdoors: alternating colors, alternating themes, etc. The upcoming photograph renders a sunset theme with alternating clouds' miens and with different shades of silver-pink in the Gulf of Riga in Latvia (Figure 2.28):

FIGURE 2.28   Alternating clouds and sunset colors in the Gulf of Riga.

Alternating colors (green and white) and themes (snowy crests and meandering green gorges) can be discerned in the alpine terrain as illustrated below (Figure 2.29):

FIGURE 2.29   Rockies' varying alpine patterns and colors.

Now we will advance with our elaborate study of alternating patterns that arise in **piecewise functions**.

2.5.1 Alternating Piecewise Functions

We encountered several examples that categorize alternating patterns in Chapter 1 and alteration between two horizontal lines $y = 1$ and $y = -1$ in Example (1.6), an alteration between diagonal lines with a positive slope and diagonal lines with a negative slope in Example (1.7), an alteration from a horizontal line and a diagonal line as in Example (2.4) in this chapter and other analogous examples. The upcoming example will render an alternating pattern between four horizontal lines.

Example 2.17. *The following sketch depicts an alteration between four horizontal lines (blue, green, red and black) on restricted intervals with length 1 as shown below:*

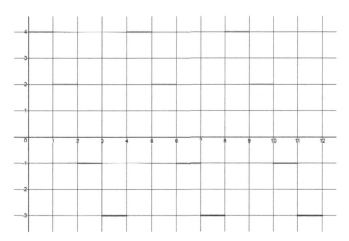

FIGURE 2.30   Piecewise function as a system of four horizontal lines.

Notice that Figure (2.30) mimics a period-4 pattern. Writing a piecewise equation that describes Figure (2.30) will be left as an end-of-chapter exercise.

The next example will illustrate a switch from a horizontal line to a diagonal line with a negative slope and so on.

Example 2.18. *The diagram below renders a specific shift starting with the blue horizontal line $y = 12$ on the interval $[0,2]$ to a green diagonal line with a negative slope on the interval $[2,5]$ and so on:*

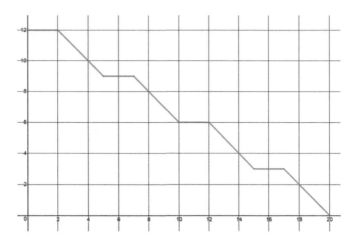

FIGURE 2.31 System of horizontal lines and diagonal lines with a negative slope.

In Figure (2.31), the blue horizontal lines are on restricted intervals with length 2, and the green diagonal lines (with a negative slope) are on restricted intervals with length 3. Writing an equation of the corresponding piecewise function in Figure (2.31) will be left as an end-of-chapter exercise.

The final example of this section will evoke a switch between two diagonal lines with slopes of opposite signs and a horizontal line.

Example 2.19. *The following scheme portrays a shift from a green diagonal line with a positive slope to a blue horizontal line and then to a red diagonal line with a negative slope as shown below:*

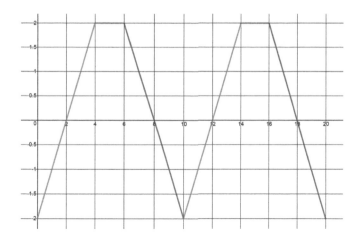

FIGURE 2.32 System of horizontal lines and diagonal lines with a positive and negative slope.

In Figure (2.32), the green diagonal lines and red diagonal lines are on restricted intervals with length 4, while the blue horizontal lines are on restricted intervals with length 2 and portrays a period-3 pattern. Writing an equation of corresponding piecewise function will be left as an end-of-chapter exercise.

2.5.2 Alternating Geometrical Patterns

In previous sections of this chapter we encountered several specimens of alternating geometric patterns at the same scale or at different scales. Example (2.9) portrays alternating right triangles at the same scale. On the other hand, Example (2.10) renders an alternating formation of diminishing squares, Example (2.12) depicts a system of alternating diminishing equilateral triangles and Example (2.14) depicts a shift between a square and a circle. Sometimes we can write one distinct sequence describing a specific pattern, and at other times, it requires a **piecewise sequence** similar to:

$$(-1)^n = \begin{cases} 1 & \text{if } n \text{ is even,} \\ -1 & \text{if } n \text{ is odd.} \end{cases}$$

We observed this alteration in Example (1.6) and similar phenomena in Example (1.7). Figure (2.15) in Example (2.9) characterizes the formation of green and blue $45-45-90$ triangles. We will revisit some of examples from the previous sections of this chapter and examine the alternating patterns in further detail and express them as **piecewise sequences** that consist of two sequences. The succeeding example will evince an alternating system of diminishing squares that we analyzed in Example (2.10) and express the sides and the areas of the squares as a **piecewise sequence**.

Example 2.20. *Recall that Figure (2.19) in Example (2.10):*

outlines the configuration of green and blue squares starting with the principal green square with length 1. By applying the Pythagorean theorem together with the identities of 45−45−90 triangles, the length of the neighboring blue square becomes $\frac{1}{\sqrt{2}}$. Hence for all $n \geq 0$, we procure the following **geometric sequence** *that traces the lengths of the squares:*

$$1, \; \frac{1}{\sqrt{2}}, \; \frac{1}{2}, \; \frac{1}{2\sqrt{2}}, \; \frac{1}{4}, \; \frac{1}{4\sqrt{2}}, \; \cdots, \; \left(\frac{1}{\sqrt{2}}\right)^{n}, \; \cdots \tag{2.9}$$

We can reformulate Eq. (2.9) as the following **piecewise sequence** *that describes the lengths of the green and blue squares:*

$$\{a_n\}_{n=0}^{\infty} = \begin{cases} \left(\frac{1}{2}\right)^{\frac{n}{2}} & \text{if } n \text{ is even,} \\ \left(\frac{1}{\sqrt{2}}\right)^{n} & \text{if } n \text{ is odd.} \end{cases} \tag{2.10}$$

The green terms of Eq. (2.10) render the length of each green square, while the blue terms of the equation characterize the length of each blue *square. Recall that the area of the principal green square is 1 and the area of the adjacent blue square is $\frac{1}{2}$. Thus for all $n \geq 0$, we procure the following* **geometric sequence** *that describes the sequence of the related areas:*

$$1, \frac{1}{2}, \frac{1}{4}, \frac{1}{8}, \frac{1}{16}, \frac{1}{32}, \cdots, \left(\frac{1}{2}\right)^{n}, \cdots \qquad (2.11)$$

We can reformulate Eq. (2.11) as the corresponding **piecewise sequence** *that describes the areas of the green and blue squares:*

$$\{A_n\}_{n=0}^{\infty} = \begin{cases} \left(\frac{1}{2}\right)^{n} & \text{if } n \text{ is even,} \\ \left(\frac{1}{2}\right)^{n} & \text{if } n \text{ is odd.} \end{cases}$$

The next example will decipher an alternating system of decreasing equilateral triangles that we analyzed in Example (2.12) and expresses the sides and areas of the triangles as a **piecewise sequence**.

Example 2.21. *Figure (2.21) in Example (2.12)*

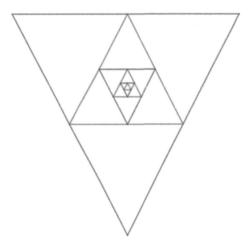

outlines the configuration of equilateral triangles starting with primary blue triangle of sides with length 2. By applying the Pythagorean theorem together with the features of 30−60−90 triangles, then the sides of *the neighboring green triangle are of length 1. Thus for all $n \geq 0$, we acquire the corresponding* **geometric sequence** *that traces the length of the triangles' sides:*

$$2, \ 1, \ \frac{1}{2}, \ \frac{1}{2^2}, \ \frac{1}{2^3}, \ \frac{1}{2^4}, \ \cdots, \ \left(\frac{1}{2}\right)^{n-1}, \ \cdots \tag{2.12}$$

We can alternatively rewrite Eq. (2.12) as the following **piecewise sequence** *that characterizes the lengths of the sides of the blue and green equilateral triangles:*

$$\{a_n\}_{n=0}^{\infty} = \begin{cases} \left(\frac{1}{2}\right)^{n-1} & \text{if } n \text{ is even,} \\ \left(\frac{1}{2}\right)^{n-1} & \text{if } n \text{ is odd.} \end{cases} \tag{2.13}$$

The blue terms of Eq. (2.13) render the length of each blue triangle's side, while the green terms of the equation depict the length of each green triangle's side. The area of the primary blue triangle is $\sqrt{3}$ and the area of the adjacent green triangle is $\frac{\sqrt{3}}{4}$. Thus for all $n \geq 0$, we acquire the following **geometric sequence** *that describes the sequence of the related areas:*

$$\sqrt{3}, \ \frac{\sqrt{3}}{4}, \ \frac{\sqrt{3}}{4^2}, \ \frac{\sqrt{3}}{4^3}, \ \frac{\sqrt{3}}{4^4}, \ \frac{\sqrt{3}}{4^5}, \ \cdots, \ \frac{\sqrt{3}}{4^n}, \ \cdots \tag{2.14}$$

We restructure Eq. (2.14) as the following piecewise sequence that describes the areas of the blue and green equilateral triangles:

$$\{A_n\}_{n=0}^{\infty} = \begin{cases} \dfrac{\sqrt{3}}{4^n} & \text{if } n \text{ is even,} \\ \dfrac{\sqrt{3}}{4^n} & \text{if } n \text{ is odd.} \end{cases}$$

The consequent example will portray an alternating system of diminishing squares and circles that we analyzed in Example (2.14) and expresses the lengths, radii and the areas as a **piecewise sequence**.

Example 2.22. *Figure (2.23) in Example (2.14) commences with inscribing a green circle inside a blue square with length 2 as illustrated below:*

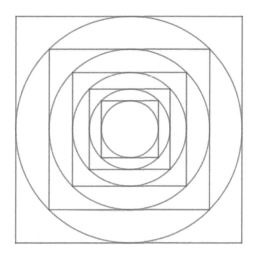

By applying the Pythagorean theorem along with the properties of 45−45−90 triangles and for all $k \geq 0$, we obtain the following **geometric sequence** *depicting the lengths of the blue squares:*

$$2, \sqrt{2}, 1, \frac{1}{\sqrt{2}}, \frac{1}{2}, \frac{1}{2\sqrt{2}}, \cdots, \left(\frac{1}{\sqrt{2}}\right)^{k-2}, \cdots, \tag{2.15}$$

and the corresponding **geometric sequence** *rendering the radii of the green circles:*

$$1, \frac{1}{\sqrt{2}}, \frac{1}{2}, \frac{1}{2\sqrt{2}}, \cdots, \left(\frac{1}{\sqrt{2}}\right)^{k}, \cdots. \tag{2.16}$$

By combining Eqs. (2.15) and (2.16), we procure the relevant **piecewise sequence:**

$$\{a_n\}_{n=0}^{\infty} = \begin{cases} \left(\frac{1}{\sqrt{2}}\right)^{\frac{n-4}{2}} & \text{if } n \text{ is even,} \\ \left(\frac{1}{\sqrt{2}}\right)^{\frac{n-1}{2}} & \text{if } n \text{ is odd.} \end{cases}$$

In addition, for all $n \geq 0$ we acquired the following **geometric sequence** *describing the areas of the blue squares:*

$$4, 2, 1, \frac{1}{2}, \cdots, \left(\frac{1}{2}\right)^{n-2}, \cdots, \tag{2.17}$$

and the following **geometric sequence** *tracing the areas of the green circles:*

$$\pi, \frac{\pi}{2}, \frac{\pi}{4}, \frac{\pi}{8}, \cdots, \frac{\pi}{2^n}, \cdots \qquad (2.18)$$

Coalescing Eqs. (2.17) and (2.18) describes the alteration from the area of a blue square to the area of a green circle and, thus, the following **piecewise sequence**:

$$\{A_n\}_{n=0}^{\infty} = \begin{cases} \left(\dfrac{1}{2}\right)^{\frac{n-4}{2}} & \text{if } n \text{ is even,} \\[2ex] \pi \left(\dfrac{1}{2}\right)^{\frac{n-1}{2}} & \text{if } n \text{ is odd.} \end{cases}$$

2.6 CHAPTER 2 EXERCISES

In problems 1–8:

1: Write a **piecewise formula** of the function in Figure (2.10).

2: Write a **piecewise formula** of the function in Figure (2.11).

3: Write a **piecewise formula** of the function in Figure (2.12).

4: Write a **piecewise formula** of the function in Figure (2.30).

5: Write a **piecewise formula** of the function in Figure (2.31).

6: Write a **piecewise formula** of the function in Figure (2.32).

7: Write a **piecewise formula** of the following function:

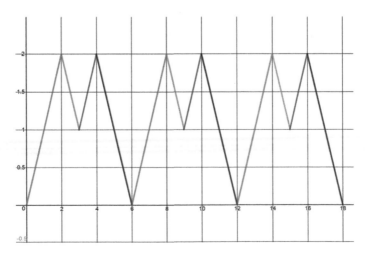

8: Write a **piecewise formula** of the following function:

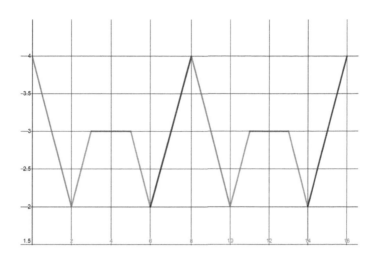

In problems 9–15:

9: Write **piecewise functions** that portray the system of shrinking rectangles:

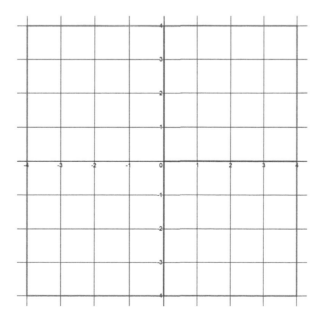

10: Write **piecewise functions** that portray the system of shrinking rectangles:

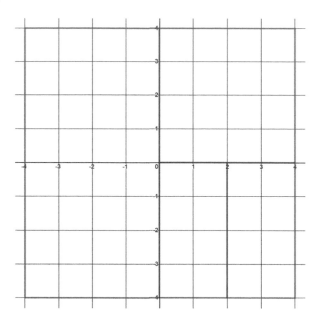

11: Write **piecewise functions** that portray the system of shrinking rectangles:

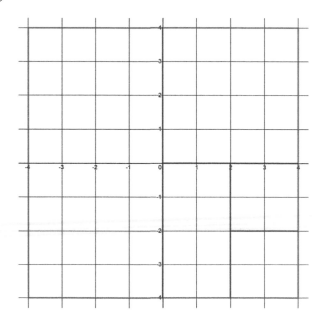

12: Write **piecewise functions** that portray the system of shrinking rectangles:

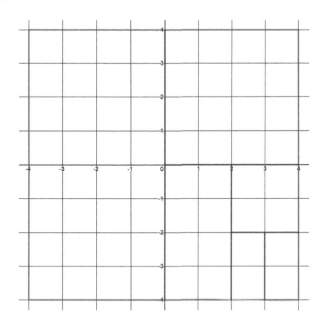

13: Write **piecewise functions** that portray the system of shrinking rectangles:

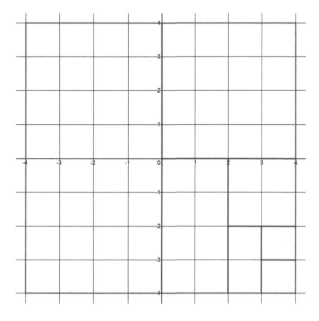

14: Write **piecewise functions** that portray the system of shrinking rectangles:

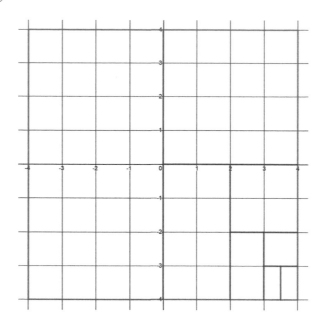

15: Write **piecewise functions** that portray the system of shrinking rectangles:

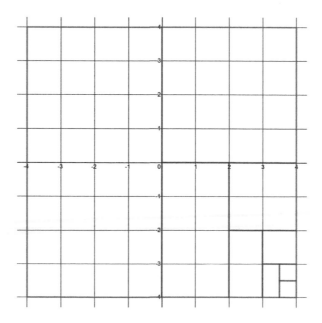

In problems 16–22:

16: Write **piecewise functions** that render the system of right triangles:

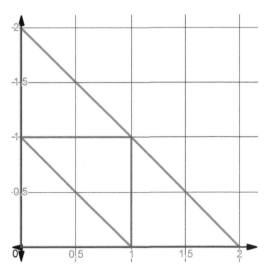

17: Write **piecewise functions** that render the system of right triangles:

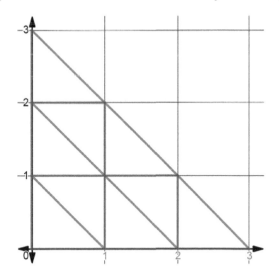

18: Write **piecewise functions** that render the system of right triangles:

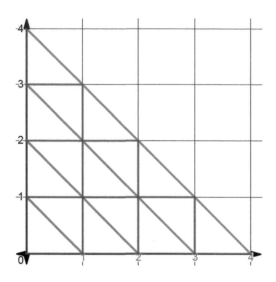

19: Write **piecewise functions** that render the system of right triangles:

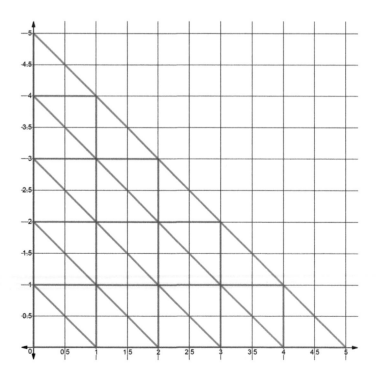

20: Write **piecewise functions** that render the system of equilateral triangles:

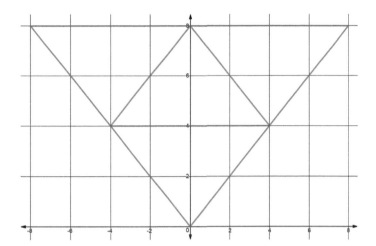

21: Write **piecewise functions** that render the system of equilateral triangles:

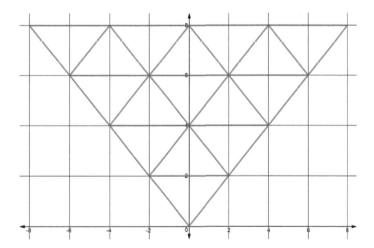

22: Write **piecewise functions** that render the system of equilateral triangles:

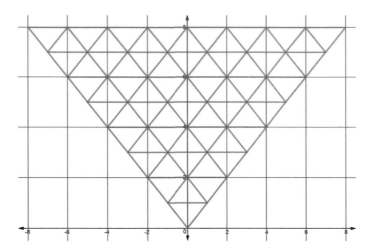

In problems 23–34:

23: Write **piecewise functions** that depict the system of shrinking equilateral triangles:

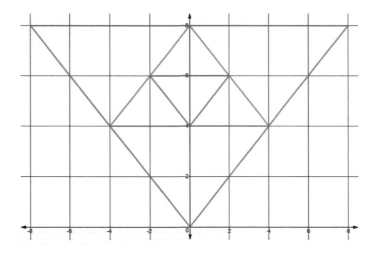

24: Write **piecewise functions** that depict the system of shrinking equilateral triangles:

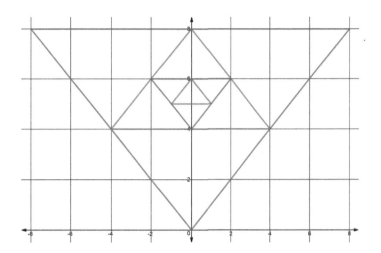

25: Write **piecewise functions** that characterize the system of shrinking squares:

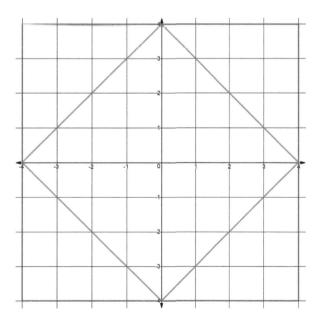

26: Write **piecewise functions** that characterize the system of shrinking squares:

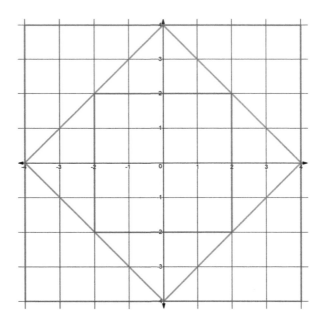

27: Write **piecewise functions** that characterize the system of shrinking squares:

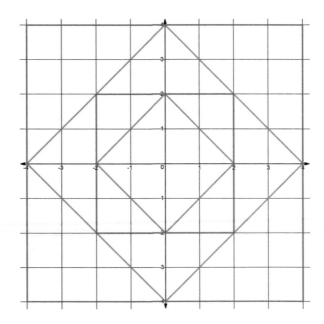

28: Write **piecewise functions** that characterize the system of shrinking squares:

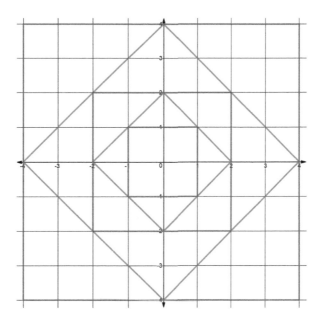

29: Write **piecewise functions** that describe the system of shrinking right triangles:

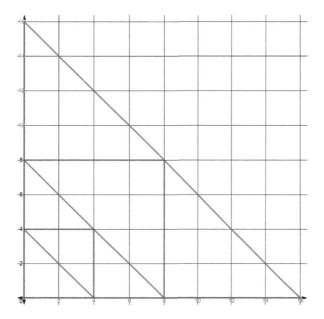

30: Write **piecewise functions** that describe the system of shrinking right triangles:

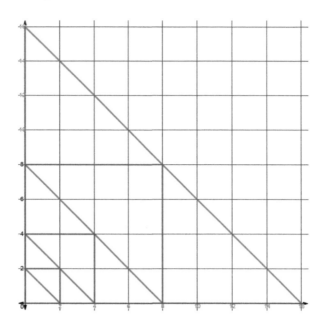

31: Write **piecewise functions** that describe the system of shrinking right triangles:

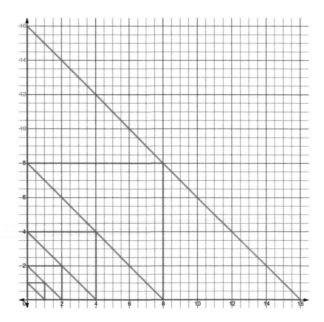

32: Write **piecewise functions** that describe the system of shrinking right triangles:

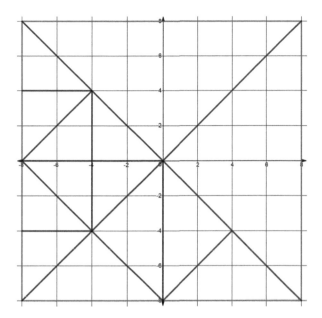

33: Write **piecewise functions** that describe the system of shrinking right triangles:

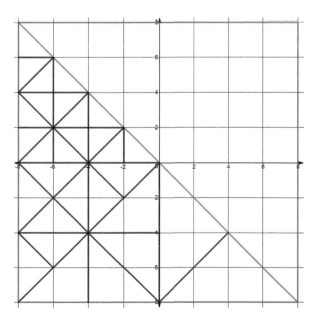

34: Write **piecewise functions** that describe the system of shrinking right triangles:

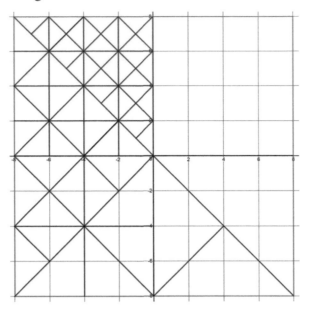

In problems 35–36:

35: Write a **formula** that determines the number of inscribed blue triangles inside the main triangle:

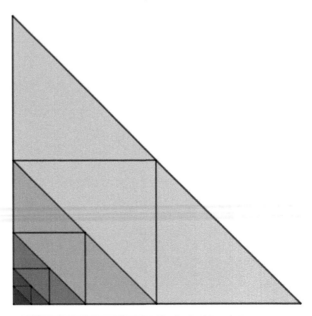

36: Write a **formula** that determines the number of inscribed blue triangles inside the main triangle:

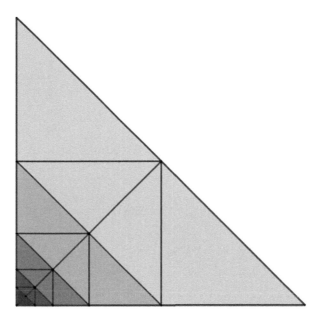

CHAPTER 3

Sequences and Summations

I**N THE PREVIOUS CHAPTERS** we encountered several sequences and patterns (linear sequences and geometric sequences) and summations. This chapter's aims are to analyze these sequences and patterns in more thorough detail, to study supplemental sequences and patterns and to get acquainted with the **Proof By Induction** method. For instance, in Figure (2.19) in Example (2.10), for all $n \in \mathbb{N}$, we discerned the **Linear Sequence** that determines the number of triangles and recites the positive multiples of 4:

$$4, \ 8, \ 12, \ 16, \ 20, \ \ldots, 4n = \{4i\}_{i=1}^{n}. \tag{3.1}$$

For all $n \in \mathbb{N}$, an analogously linear pattern emerged from Figure (2.21) in Example (2.13):

$$1, \ 4, \ 7, \ 10, \ \ldots, \ 3n + 1 = \{3i + 1\}_{i=0}^{n}. \tag{3.2}$$

Our objective is to examine analogous linear sequences and determine their corresponding formulations. Furthermore, from Figure (2.19) in Example (2.10), for all $n \geq 0$ we procured the following **geometric sequence**:

$$1, \ \frac{1}{\sqrt{2}}, \ \frac{1}{2}, \ \frac{1}{2\sqrt{2}}, \ \frac{1}{4}, \ \frac{1}{4\sqrt{2}}, \ \ldots, \ \left(\frac{1}{\sqrt{2}}\right)^{n} = \left\{\left(\frac{1}{\sqrt{2}}\right)^{i}\right\}_{i=0}^{n}. \tag{3.3}$$

Notice that Eqs. (3.1)–(3.3) are finite sequences that have a starting and terminating index as stated in Definition (1). We will also examine infinite sequences in the form $\{x_n\}_{n=1}^{\infty}$ that have only a starting index as stated in Definition (2). Our goals are to meticulously explore supplemental geometric sequences and their structure and factorial-type sequences that mimic geometric sequences. Moreover, we will examine the **Factorial Pattern** in comparison to the Linear and Geometric Sequences and will investigate additional **Factorial-Type Sequences**, **Piecewise Sequences** and **Alternating Sequences**.

For all $n \in \mathbb{N}$, Figure (2.14) in Example (2.7) rendered the corresponding summation of the consecutive positive integers:

$$1 + 2 + 3 + 4 + \ldots + (n-1) + n = \sum_{i=1}^{n} i = \frac{n \cdot [n+1]}{2}. \qquad (3.4)$$

In addition, for all $n \in \mathbb{N}$, Figure (2.15) in Example (2.8) indicated the related sum of the consecutive positive odd integers:

$$1 + 3 + 5 + \ldots + (2n-3) + (2n-1) = \sum_{i=1}^{n} (2i-1) = n^2. \qquad (3.5)$$

Furthermore, for all $n \in \mathbb{N}$, Figure (2.24) in Example (2.15) describes the corresponding **Geometric Summation** (provided that $r \neq 1$):

$$a + a \cdot r + a \cdot r^2 + a \cdot r^3 + \ldots + a \cdot r^n = \sum_{i=0}^{n} a \cdot r^i = \frac{a \cdot [1 - r^{n+1}]}{1 - r}. \qquad (3.6)$$

We will prove Eqs. (3.4)–(3.6) by using the **Proof By Induction** technique and derive and prove supplemental summations.

3.1 LINEAR AND QUADRATIC SEQUENCES

A **Linear Sequence** is assembled by adding a constant from neighbor to neighbor as we observed in various examples in Chapters 1 and 2. Our aim is to write a formula that recites all the specified terms of the given sequence with a starting term and an ending term and the analogous starting and ending indices. We will examine examples with a range of patterns. The succeeding example adds a six while shifting from neighbor to neighbor.

Example 3.1. *Write a formula of the following sequence:*

$$11, \ 17, \ 23, \ 29, \ 35, \ \ldots, \ 149. \tag{3.7}$$

Solution: *The starting term of Eq. (3.7) is 11 and the ending term of the equation is 149; both are 1 less than the multiples of 6. To determine the starting and terminating indices we set*

$$6i - 1 = 11 \quad \text{and} \quad 6i - 1 = 149.$$

We then obtain the corresponding starting index $i = 2$, the terminating index $i = 25$ and the corresponding formula depicting the 24 terms of (3.7):

$$\{6i - 1\}_{i=2}^{25}. \tag{3.8}$$

Alternatively, we can rewrite Eq. (3.8) by shifting the starting and ending indices down by 1 and acquire:

$$\{6i + 5\}_{i=1}^{24}.$$

The next two examples will focus on quadratic sequences that describe the perfect squares.

Example 3.2. *Write a formula of the following sequence:*

$$9, \ 16, \ 25, \ 36, \ 49, \ 64, \ \ldots, \ 1600. \tag{3.9}$$

Solution: *Eq. (3.9) starts at 9 and terminates at 1600. By setting*

$$i^2 = 9 \quad \text{and} \quad i^2 = 1600,$$

we obtain $i = 3$ and $i = 40$ and the corresponding formula listing the 38 terms of Eq. (3.9):

$$\{i^2\}_{i=3}^{40}. \tag{3.10}$$

Similar to Example (5.1), we rewrite Eq. (3.10) by shifting the indices down by 2 and obtain:

$$\{(i + 2)^2\}_{i=1}^{38}.$$

The upcoming example evokes a sequence that describes the consecutive even perfect squares starting at 4.

Example 3.3. *Write a formula of the following sequence:*

$$4, \ 16, \ 36, \ 64, \ 100, \ 144, \ \ldots, \ 3600. \tag{3.11}$$

Solution: *Observe that Eq. (3.11) starts at 1 and terminates at 3600. By setting*

$$4i^2 = 4 \quad \text{and} \quad 4i^2 = 3600,$$

we acquire $i = 1$, $i = 30$ and the comparable formula of (3.11):

$$\{4i^2\}_{i=1}^{30}.$$

The upcoming sub-section will focus on the **Summation-Type** sequences.

3.1.1 Summation-Type Sequences

The consequent two examples resemble a **Summation-Type** sequence analogous to (3.4) and (3.5). In the first example we will add the consecutive positive odd integers starting at 1 (adding a 3 to the first neighbor, adding a 5 to the next neighbor, etc.)

Example 3.4. *Write a formula of the following sequence:*

$$1, \ 4, \ 9, \ 16, \ 25, \ 36, \ 49, \ldots. \tag{3.12}$$

Solution: *Note that Eq. (3.12) commences with 1, we add a 3 to transition to the next neighbor and then add a 5 to the consequent neighbor and so on. We will denote $n \in \mathbb{N}$ as the terminating index. Notice:*

1,

$1 + 3 = 4,$

$4 + 5 = [1 + 3] + 5 = 9,$

$9 + 7 = [1 + 3 + 5] + 7 = 16,$

$16 + 9 = [1 + 3 + 5 + 7] + 9 = 25,$

$25 + 11 = [1 + 3 + 5 + 7 + 9] + 11 = 36,$

$\vdots$

By induction, for all $n \in \mathbb{N}$ we acquire:

$$\{a_i\}_{i=0}^{n-1} = \left[\sum_{i=0}^{n-1} (2i + 1)\right]. \tag{3.13}$$

We can reformulate Eq. (3.13) by shifting the starting index up by 1 and procure:

$$\{a_i\}_{i=1}^{n} = \left[\sum_{i=1}^{n} (2i - 1) \right]. \qquad (3.14)$$

We encountered Eq. (3.14) in Example (2.7) and will study additional analogous summations in the later section of this chapter. In Chapter 5, we will write an alternative recursive formula of the assigned sequence.

Example 3.5. *Write a formula of the following sequence:*

$$1, \ 2, \ 4, \ 7, \ 11, \ 16, \ 22, \dots. \qquad (3.15)$$

Solution: *The starting term of Eq. (3.15) is 1; we add a 1 to transition to the next neighbor; and then we add a 2 to the consequent neighbor and so on. Let $n \in \mathbb{N}$ be the terminating index. Observe:*

$$1,$$
$$1 + [1] = 2,$$
$$1 + [1 + 2] = 4,$$
$$1 + [1 + 2 + 3] = 7,$$
$$1 + [1 + 2 + 3 + 4] = 11,$$
$$1 + [1 + 2 + 3 + 4 + 5] = 16,$$
$$\vdots$$

By induction, for all $n \geq 0$ we obtain:

$$\{a_i\}_{i=0}^{n} = 1 + \left[\sum_{i=0}^{n} i \right]. \qquad (3.16)$$

We can rewrite Eq. (3.16) by shifting the index up by 1 and get:

$$\{a_i\}_{i=1}^{n+1} = 1 + \left[\sum_{i=1}^{n+1} (i - 1) \right].$$

Similar problems will be provided as end-of-chapter exercises. Now we will transition to **geometric** and **factorial-type** sequences.

3.2 GEOMETRIC AND FACTORIAL-TYPE SEQUENCES

A **geometric sequence** is contrived by multiplying by a constant from term to term as we observed in various examples in the preceding chapters. For instance, in Example (1.9), for $n \geq 0$ we defined the finite **geometric sequence** with $n + 1$ terms in the form:

$$\{a \cdot r^i\}_{i=0}^n = a, \ a \cdot r, \ a \cdot r^2, \ a \cdot r^3, \ a \cdot r^4, \ \ldots, \ a \cdot r^n, \qquad (3.17)$$

where a is the starting term of Eq. (3.17) and r is the multiplicative factor. We will commence with repetitive-type examples that will illustrate the change of indices and render infinite sequences with a starting index only.

Example 3.6. *Write a formula of the following sequence:*

$$3, \ 12, \ 48, \ 192, \ 768, \ \ldots.$$

Solution: *Observe:*

$$3,$$
$$3 \cdot 4 = 12,$$
$$12 \cdot 4 = 3 \cdot 4^2 = 48,$$
$$48 \cdot 4 = 3 \cdot 4^3 = 192,$$
$$192 \cdot 4 = 3 \cdot 4^4 = 768,$$
$$\vdots$$

For all $n \geq 0$ we procure:

$$\{3 \cdot 4^n\}_{n=0}^\infty. \qquad (3.18)$$

By rearranging the index by 1, we rewrite Eq. (3.18) as:

$$\{3 \cdot 4^{n-1}\}_{n=1}^\infty.$$

The upcoming example will mimic the techniques in Examples (2.10) and (2.14).

Example 3.7. *Write a formula of the following sequence:*

$$9, \ 3\sqrt{3}, \ 3, \ \sqrt{3}, \ 1, \ \frac{1}{\sqrt{3}}, \ \ldots.$$

Solution: *Notice:*

$$9,$$

$$\frac{9}{\sqrt{3}} = 3\sqrt{3},$$

$$\frac{3\sqrt{3}}{\sqrt{3}} = 3,$$

$$\frac{3}{\sqrt{3}} = \sqrt{3},$$

$$\frac{\sqrt{3}}{\sqrt{3}} = 1,$$

$$\vdots$$

Hence for all n ≥ 0 we acquire:

$$\left\{ \frac{9}{(\sqrt{3})^n} \right\}_{n=0}^{\infty} . \tag{3.19}$$

Alternatively we can recast Eq. (3.19) as:

$$\left\{ \frac{1}{(\sqrt{3})^{n-4}} \right\}_{n=0}^{\infty} .$$

The upcoming sub-section will resume with the **Factorial-Type** Patterns.

3.2.1 Factorial-Type Sequences

Analogous to Example (3.4), the upcoming example evokes the geometric-type pattern that guides us to the **factorial-type** sequences.

Example 3.8. *Write a formula of the following sequence:*

$$3, \ 6, \ 18, \ 72, \ 360, \ 2160, \ \ldots .$$

Solution: *Note:*

$$3 \cdot [\mathbf{1}] = 3,$$

$$3 \cdot [\mathbf{1} \cdot \mathbf{2}] = 6,$$

$$6 \cdot 3 = 3 \cdot [\mathbf{2} \cdot \mathbf{3}] = 18,$$

$$18 \cdot 4 = 3 \cdot [\mathbf{2} \cdot \mathbf{3} \cdot \mathbf{4}] = 72,$$

$$72 \cdot 5 = 3 \cdot [\mathbf{2} \cdot \mathbf{3} \cdot \mathbf{4} \cdot \mathbf{5}] = 360,$$

$$\vdots$$

Parallel to the techniques in Example (3.4), for all $n \in \mathbb{N}$ we procure:

$$\{x_n\}_{i=1}^n = 3 \cdot \left[\prod_{i=1}^n i\right].$$

We will resume with our inquiries on the **Factorial** and the **Factorial-Type** Patterns. In Chapter 1, for all $n \in \mathbb{N}$ we produced the **Factorial** by multiplying the consecutive positive integers starting at 1 as:

$$0! = 1 \quad \text{and} \quad n! = \prod_{i=1}^n i. \tag{3.20}$$

Note that Eq. (3.20) can be expressed as the following **piecewise sequence**:

$$n! = \begin{cases} 1 & \text{if } n = 0, \\ \prod_{i=1}^n i & \text{if } n \in \mathbb{N}. \end{cases}$$

In Chapter 5 we will rewrite the Factorial as a recursive sequence. Now we will transition to assorted examples of patterns that mimic the **Factorial Pattern**. The upcoming example will render the product of the consecutive positive even integers.

Example 3.9. *Write a formula of the following sequence:*

$$2, \ 8, \ 48, \ 384, \ 3840, \ \ldots.$$

Solution: *Observe:*

$$2,$$
$$[2] \cdot 4 = 8,$$
$$8 \cdot 6 = [2 \cdot 4] \cdot 6 = 48,$$
$$48 \cdot 8 = [2 \cdot 4 \cdot 6] \cdot 8 = 384,$$
$$384 \cdot 10 = [2 \cdot 4 \cdot 6 \cdot 8] \cdot 10 = 3840,$$
$$\vdots$$

Thus, for all $n \in \mathbb{N}$ we acquire:

$$\{x_n\}_{i=1}^n = \prod_{i=1}^n 2i. \tag{3.21}$$

We can also express Eq. (3.21) as a product of two sequences:

$$\{x_n\}_{i=1}^n = \{2^i\}_{i=1}^n \cdot \left[\prod_{i=1}^n i\right] = 2^n \cdot n!.$$

The succeeding example will require utilizing a **piecewise sequence** to characterize the assigned sequence.

Example 3.10. *Write a formula of the following sequence:*

$$2, \ 10, \ 70, \ 630, \ 6930, \ \ldots.$$

Solution: *Notice:*

$$2,$$
$$2 \cdot [5] = 10,$$
$$10 \cdot 7 = 2 \cdot [5 \cdot 7] = 70,$$
$$70 \cdot 9 = 2 \cdot [5 \cdot 7 \cdot 9] = 630,$$
$$630 \cdot 11 = 2 \cdot [5 \cdot 7 \cdot 9 \cdot 11] = 6930,$$
$$\vdots$$

Then for all $n \in \mathbb{N}$, we obtain the following **piecewise sequence***:*

$$\{x_n\}_{i=1}^n = \begin{cases} 2 & \text{if } n = 1, \\ 2 \cdot \left[\prod_{i=2}^n (2i+1)\right] & \text{if } n \geq 2. \end{cases}$$

Observe that the product formula works only starting with the second term of the sequence and, therefore, requires the use of a piecewise formula.

3.3 ALTERNATING AND PIECEWISE SEQUENCES

In Examples (1.6) and (2.9), we detected the following **alternating piecewise sequence** that diverts between 1 and -1. In fact for all $n \geq 0$:

$$\{(-1)^n\}_{n=0}^\infty = \begin{cases} 1 & \text{if } n \text{ is even}, \\ -1 & \text{if } n \text{ is odd}. \end{cases}$$

Furthermore, we encountered the alternating geometric sequences in Examples (2.20)–(2.22). The section's aims are to get acquainted with assorted alternating and piecewise sequences. Alternating sequences can be written in one fragment as the sequence above. However, several piecewise sequences cannot be written in one fragment as we experienced in Example (3.10) in the previous section. We will distinguish the similarities and differences after several repetitive-type examples. The upcoming example will manifest an alternating geometric sequence.

Example 3.11. *Write a formula of the following sequence:*

$$5, \ -10, \ 20, \ -40, \ 80, \ \ldots. \tag{3.22}$$

Solution: *First of all, note that the first term of Eq. (3.22) is positive and the sign switches from neighbor to neighbor. Second of all, Eq. (3.22) is a geometric sequence with $a = 5$ and $r = -2$. Hence for $n \geq 0$ we get:*

$$\{x_n\}_{n=0}^{\infty} = \{5 \cdot (-2)^n\}_{n=0}^{\infty} = \{5 \cdot 2^n \cdot (-1)^n\}_{n=0}^{\infty}. \tag{3.23}$$

Now we rewrite Eq. (3.22) as:

$$5, \ -10, \ 20, \ -40, \ 80, \ \ldots. \tag{3.24}$$

Thus we express Eq. (3.24) as the following **piecewise geometric sequence***:*

$$\{x_n\}_{n=0}^{\infty} = \begin{cases} \{5 \cdot 2^n\}_{n=0}^{\infty} & \text{if } n \text{ is even,} \\ \{-10 \cdot 2^{2n-2}\}_{n=1}^{\infty} & \text{if } n \text{ is odd.} \end{cases}$$

Example (3.11) extends to the following **Alternating Geometric Sequences**:

$$a, \ -a \cdot r, \ a \cdot r^2, \ -a \cdot r^3, \ a \cdot r^4, \ \ldots = \{a \cdot r^n \cdot (-1)^n\}_{n=0}^{\infty},$$

and

$$-a, \ a \cdot r, \ -a \cdot r^2, \ a \cdot r^3, \ -a \cdot r^4, \ \ldots = \{a \cdot r^n \cdot (-1)^{n+1}\}_{n=0}^{\infty}.$$

Similar to Example (2.10), the upcoming example will exhibit an alternating linear sequence.

Example 3.12. *Write a formula of the following sequence:*

$$-6, \ 12, \ -18, \ 24, \ -30, \ 36, \ \ldots. \tag{3.25}$$

Solution: *Note that Eq. (3.25) lists multiples of 6 starting with a negative term, then the sign switches to positive and so on. Thus, for all $n \in \mathbb{N}$ we procure:*

$$\{x_n\}_{n=1}^{\infty} = \{(-1)^n \, 6n\}_{n=1}^{\infty}. \tag{3.26}$$

Now we revise Eq. (3.25) as:

$$-6, \; 12, \; -18, \; 24, \; -30, \; 36, \; \dots. \tag{3.27}$$

Thus we evoke Eq. (3.27) as the following **piecewise linear sequence***:*

$$\{x_n\}_{n=0}^{\infty} = \begin{cases} -[6(n+1)] & \text{if } n \text{ is even,} \\ 6(n+1) & \text{if } n \text{ is odd.} \end{cases}$$

The consequent example will require decomposition into two sub-sequences as every fourth term of the sequence alternates in sign. In fact, we will discover an alternating sub-sequence and a nonalternating sub-sequence.

Example 3.13. *Write a formula of the following sequence:*

$$2, \; 4, \; 6, \; -8, \; 10, \; 12, \; 14, \; -16, \; \dots. \tag{3.28}$$

Solution: *Note that Eq. (3.28) is composed in terms of positive even integers starting at 2 while every fourth term of (3.28) is negative. This is quite different to what we encountered in Examples (3.11) and (3.12). Thus, we will break up (3.28) into two primary blue and green sub-sequences:*

$$2, \; 4, \; 6, \; -8, \; 10, \; 12, \; 14, \; -16, \; \dots. \tag{3.29}$$

Now observe that in (3.29) the blue sub-sequence is a nonalternating sequence while the green sub-sequence alternates. Hence for $n \geq 0$ we acquire:

$$\{x_n\}_{n=0}^{\infty} = \begin{cases} 2(n+1) & \text{if } n \text{ is even,} \\ (-1)^{\frac{n-1}{2}}[2(n+1)] & \text{if } n \text{ is odd.} \end{cases}$$

3.4 SUMMATIONS AND PROOF BY INDUCTION

This section's aims are to study summations similar to Eqs. (3.4)–(3.6) and prove them by applying the **Proof By Induction** method. We will also derive further summations from Eqs. (3.4)–(3.6). The next example will illustrate the **Proof By Induction** method to prove Eq. (3.4).

Example 3.14. *Using* **Proof By Induction**, *verify the following summation:*

$$1 + 2 + 3 + 4 + \ldots + (n - 1) + n = \sum_{i=1}^{n} i = \frac{n \cdot [n + 1]}{2}. \qquad (3.30)$$

Solution: *Observe that Eq. (3.30) holds true for n = 2 as we get:*

$$1 + 2 = \frac{2 \cdot 3}{2} = 3.$$

Now we will assume that Eq. (3.30) holds true for n = k:

$$1 + 2 + 3 + 4 + \ldots + (k - 1) + k = \sum_{i=1}^{k} i = \frac{k \cdot [k + 1]}{2}. \qquad (3.31)$$

Next we will verify that Eq. (3.31) holds true for n = k + 1:

$$[1 + 2 + 3 + 4 + \ldots + k] + [k + 1] = \sum_{i=1}^{k+1} i = \frac{[k + 1] \cdot [k + 2]}{2}. \qquad (3.32)$$

Using Eq. (3.31) we reformulate Eq. (3.32) as:

$$\left[\sum_{i=1}^{k} i \right] + [k + 1] = \left[\frac{k \cdot [k + 1]}{2} \right] + [k + 1] = \frac{[k + 1] \cdot [k + 2]}{2}.$$

Hence, the result follows.

The consequent example will apply the **Proof By Induction** method to prove a special case of (3.6).

Example 3.15. *Using* **Proof By Induction**, *verify the following sum:*

$$1 + 2 + 2^2 + 2^3 + \ldots + 2^{n-1} + 2^n = \sum_{i=0}^{n} 2^i = 2^{n+1} - 1. \qquad (3.33)$$

Solution: *Observe that Eq. (3.33) holds true for n = 2 as we get:*

$$1 + 2 + 2^2 = 7 = 2^3 - 1.$$

Now we will assume that Eq. (3.33) holds true for n = k:

$$1 + 2 + 2^2 + 2^3 + \ldots + 2^{k-1} + 2^k = \sum_{i=0}^{k} 2^i = 2^{k+1} - 1. \qquad (3.34)$$

Now we will confirm that Eq. (3.34) holds true for $n = k + 1$:

$$\left[1 + 2 + 2^2 + 2^3 + \ldots + 2^k\right] + \left[2^{k+1}\right] = \sum_{i=0}^{k+1} 2^i = 2^{k+2} - 1. \quad (3.35)$$

Using Eq. (3.34), we restructure Eq. (3.35) as:

$$\left[\sum_{i=0}^{k} 2^i\right] + \left[2^{k+1}\right] = \left[2^{k+1} - 1\right] + 2^{k+1} = 2 \cdot 2^{k+1} - 1 = 2^{k+2} - 1.$$

Hence, the result follows.

The next sequence of examples will apply Eq. (3.4) to derive a specific summation.

Example 3.16. *Using Eq. (3.4), derive the following finite sum:*

$$4 + 8 + 12 + 16 + 20 + 24 + \ldots. \quad (3.36)$$

Solution: *Using the Sigma Notation, for $n \in \mathbb{N}$ we amend Eq. (3.36) as:*

$$4 + 8 + 12 + 16 + 20 + 24 + \ldots + 4n = \sum_{i=1}^{n} 4i. \quad (3.37)$$

Notice that Eq. (3.37) adds multiples of 4. Now we factor the 4 and obtain:

$$4 \cdot [1 + 2 + 3 + 4 + \ldots + n] = 4 \cdot \left[\sum_{i=1}^{n} i\right] = 2n \cdot [n + 1]. \quad (3.38)$$

Proving Eq. (3.38) by induction will be left as an end-of-chapter exercise.

Example 3.17. *Using Eq. (3.4), derive the formula of the corresponding sum:*

$$4 + 7 + 10 + 13 + 16 + 19 + \ldots. \quad (3.39)$$

Solution: *Using the Sigma Notation, for $n \in \mathbb{N}$ we reformulate Eq. (3.39) as:*

$$4 + 7 + 10 + 13 + 16 + 19 + \ldots + (3n + 1) = \sum_{i=1}^{n} (3i + 1). \quad (3.40)$$

*Now using the **Distributive Property** of Summations, we decompose Eq. (3.40) and acquire:*

$$\sum_{i=1}^{n} (3i+1) = \sum_{i=1}^{n} 3i + \sum_{i=1}^{n} 1 = \frac{3n \cdot [n + 1]}{2} + n = \frac{n \cdot [3n + 5]}{2}. \quad (3.41)$$

Proving Eq. (3.41) by induction will be left as an end-of-chapter exercise.

Example 3.18. *Using Eq. (3.4), derive the formula of the following summation:*

$$5 + 6 + 7 + 8 + 9 + 10 + 11 + \dots. \tag{3.42}$$

Solution: *Using the Sigma Notation, for $n \geq 5$ we compose Eq. (3.42) as:*

$$5 + 6 + 7 + 8 + 9 + 10 + 11 + \dots + n = \sum_{i=5}^{n} i. \tag{3.43}$$

Now by revising the starting and the terminating indices by 4 units, for $n \geq 5$ we obtain the corresponding summation:

$$\sum_{i=5}^{n} i = \sum_{i=1}^{n-4} (i + 4). \tag{3.44}$$

*By implementing the **Distributive Property** of Summations, we split Eq. (3.44) into two separate summations and for all $n \geq 5$ we procure:*

$$\sum_{i=1}^{n-4} (i + 4) = \sum_{i=1}^{n-4} i + \sum_{i=1}^{n-4} 1 = \frac{[n - 4] \cdot [n - 3]}{2} + [n - 4]$$
$$= \frac{[n - 4] \cdot [n - 1]}{2}.$$

From Example (3.18), for all $m \in [2, \dots, n]$ we can then extend Eq. (3.43) to the related summation of specific consecutive positive integers:

$$m + (m + 1) + (m + 2) + (m + 3) + (m + 4) + \dots + n = \sum_{i=m}^{n} i. \tag{3.45}$$

Obtaining a summation formula of Eq. (3.45) will be left as an end-of-chapter exercise.

3.5 CHAPTER 3 EXERCISES

In problems 1–12, write a **formula** of each sequence:

1: 1, 5, 9, 13, 17, 21, 25, ...

2: 4, 11, 18, 25, 32, 39, 46, ...

3: 8, 17, 26, 35, 44, 53, 62, ...

4: $(m + 8), (m + 12), (m + 16), (m + 20), (m + 24), \dots.$

5: 1, 9, 25, 49, 81, 121, 169,....

6: 1, 25, 81, 169, 289, 441, 625,....

7: 9, 49, 121, 225, 361, 529, 729,....

8: 16, 64, 144, 256, 400, 576, 784,....

9: 4, 36, 100, 196, 324, 484, 676,....

10: 2, 6, 12, 20, 30, 42, 56,....

11: 4, 7, 12, 19, 28, 39, 52,....

12: 5, 8, 14, 23, 35, 50, 68,....

In problems 13–22, write a **formula** of each sequence:

13: 2, 6, 18, 54, 162, 486, 1458,....

14: 2, $\sqrt{8}$, 4, $\sqrt{32}$, 8, $\sqrt{128}$, 16,....

15: 2, $\sqrt{12}$, 6, $\sqrt{108}$, 18, $\sqrt{972}$, 54,....

16: 162, 54, 18, 6, 2, $\frac{2}{3}$, $\frac{2}{9}$,....

17: 12, 18, 27, $\dfrac{81}{2}$, $\dfrac{243}{4}$, $\dfrac{729}{8}$, $\dfrac{2187}{16}$,....

18: 45, 30, 20, $\dfrac{40}{3}$, $\dfrac{80}{9}$, $\dfrac{160}{27}$, $\dfrac{320}{81}$,....

19: 1, 3, 15, 105, 945, 10,395, 135,135,....

20: 4, 8, 32, 192, 1536, 15,360, ...

21: 1, 3, 21, 231, 3465, 65,835, ...

22: $\dfrac{5}{16}$, $\dfrac{5}{4}$, 10, 120, 1920, 38,400, ...

23: 5, 20, 100, 600, 4200, 33,600, 302,400, ...

24: 2, 6, 30, 210, 1890, 20,790, 270,270, ...

25: 1, 2, 8, 48, 384, 3840, 46,080, ...

26: $\dfrac{1}{9}$, $\dfrac{5}{9}$, $\dfrac{35}{9}$, 35, 385, 5005, 75,075, ...

In problems 27–32, write a **formula** of each piecewise sequence:

27: $-4,\ 10,\ -16,\ 22,\ -28,\ 34,\ldots$

28: $2,\ 4,\ -6,\ 8,\ 10,\ 12,\ -14,\ 16,\ \ldots$

29: $-1,\ \sqrt{2},\ 2,\ \sqrt{8},\ -4,\ \sqrt{32},\ 8,\ \sqrt{128},\ \ldots$

30: $-9,\ -12,\ 15,\ 18,\ -21,\ -24,\ 27,\ 30,\ \ldots$

31: $5,\ 10,\ -15,\ -20,\ 25,\ 30,\ -35,\ -40,\ \ldots$

32: $6,\ -12,\ -18,\ 24,\ 30,\ -36,\ -42,\ 48,\ \ldots$

In problems 33–36 use (3.4) and the **Distributive Property** to determine the following summations:

33: $3 + 6 + 9 + 12 + 15 + \ldots + 240.$

34: $4 + 8 + 12 + 16 + 20 + \ldots + 320.$

35: $5 + 9 + 13 + 17 + 21 + \ldots + 81.$

36: $2 + 5 + 8 + 11 + 14 + \ldots + 92.$

In problems 37–40 use (3.6) to determine the following summations:

37: $\dfrac{1}{4} + \dfrac{1}{2} + 1 + 2 + 4 + \ldots + 128.$

38: $4 + 2\sqrt{2} + 2 + \sqrt{2} + 1 + \frac{1}{\sqrt{2}} + \ldots + \frac{1}{8}.$

39: $\dfrac{1}{9} - \dfrac{1}{3} + 1 - 3 + 9 - \ldots + 729.$

40: $1 - \dfrac{\sqrt{3}}{2} + \dfrac{3}{4} - \dfrac{3\sqrt{3}}{8} + \dfrac{9}{16} - \ldots + \dfrac{243}{1024}.$

In problems 41–50 prove the following expressions by the **Proof By Induction** method:

41: $\sum_{i=1}^{k} (2i - 1) = k^2$

42: $\sum_{i=1}^{k} (4i - 3) = k \cdot [2k - 1]$

43: $\sum_{i=1}^{k} i^2 = \frac{k \cdot [k+1] \cdot [2k+1]}{6}$

44: $\sum_{i=1}^{k} i[i + 1] = \frac{k \cdot [k+1] \cdot [k+2]}{3}$

45: $\sum_{i=0}^{k} a \cdot r^i = \frac{a \cdot [1-r^{k+1}]}{1-r}$, $(r \neq 1)$

46: $\sum_{i=1}^{k} i \cdot [i+1] = \frac{k \cdot [k+1] \cdot [k+2]}{3}$

47: $\sum_{i=1}^{k} \frac{1}{i \cdot [i+1]} = \frac{k}{k+1}$

48: $\sum_{i=1}^{k} \frac{1}{[2i-1] \cdot [2i+1]} = \frac{k}{2k+1}$

49: $\sum_{i=1}^{k} i \cdot 2^{i-1} = [k-1] \cdot 2^k + 1$.

50: $\sum_{i=1}^{k} i \cdot i! = (k+1)! - 1$

In problems 51–54, use (3.4) and the **Distributive Property** to derive the following summations:

51: $\sum_{i=7}^{k} i$, $(i \geq 7)$

52: $\sum_{i=8}^{k} i$, $(i \geq 8)$

53: $\sum_{i=9}^{k} i$, $(i \geq 9)$

54: Using problems 51–53 determine the formula of (3.45).

CHAPTER 4

Pascal's Triangle Identities

I N CHAPTER 1 WE CONVENED with **Pascal's Triangle**:

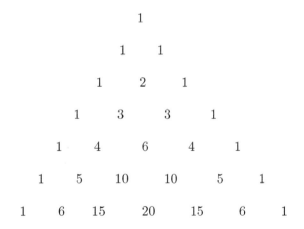

FIGURE 4.1   Seven rows of Pascal's triangle.

This chapter's aims are to derive various traits of Pascal's triangle and prove them by applying the Definition of Combinations (4.1) and by Induction. We will examine Figure (4.1) in the horizontal direction to analyze the triangle's **Horizontally Oriented Identities** and in the diagonal direction to examine the triangle's **Diagonally Oriented Identities**.

To obtain the triangle's Horizontally Oriented Identities, we will decompose Figure (4.1) into blue and red horizontal rows (the blue rows are even-ordered rows and the red rows are odd-ordered rows) as illustrated in the proceeding diagram (Figure 4.2):

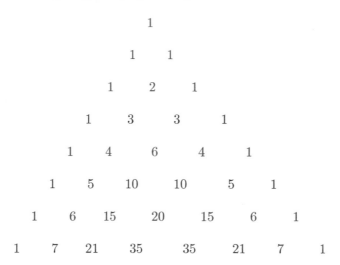

FIGURE 4.2 Pascal's triangle decomposed into blue and red rows.

Analogously, to acquire the triangle's Diagonally Oriented Identities, we will break up Figure (4.1) into blue and red diagonals as indicated below (Figure 4.3):

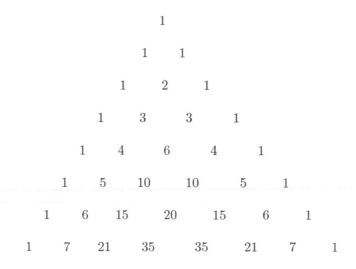

FIGURE 4.3 Pascal's triangle decomposed into blue and red diagonals.

Furthermore, using the Definition of Combinations (4.1) we can reformulate Figure (4.1) as rendered in the diagram below:

$$\binom{0}{0}$$

$$\binom{1}{0} \quad \binom{1}{1}$$

$$\binom{2}{0} \quad \binom{2}{1} \quad \binom{2}{2}$$

$$\binom{3}{0} \quad \binom{3}{1} \quad \binom{3}{2} \quad \binom{3}{3}$$

$$\binom{4}{0} \quad \binom{4}{1} \quad \binom{4}{2} \quad \binom{4}{3} \quad \binom{4}{4}$$

$$\binom{5}{0} \quad \binom{5}{1} \quad \binom{5}{2} \quad \binom{5}{3} \quad \binom{5}{4} \quad \binom{5}{5}$$

$$\binom{6}{0} \quad \binom{6}{1} \quad \binom{6}{2} \quad \binom{6}{3} \quad \binom{6}{4} \quad \binom{6}{5} \quad \binom{6}{6}$$

$$\binom{7}{0} \quad \binom{7}{1} \quad \binom{7}{2} \quad \binom{7}{3} \quad \binom{7}{4} \quad \binom{7}{5} \quad \binom{7}{6} \quad \binom{7}{7}$$

FIGURE 4.4 Pascal's triangle expressed in Combinations.

Figure (4.4) guides us to the definition of **Combinations** expressed in terms of factorials.

Definition 9. *For all $n \geq 0$ and $k \in [0, 1, \ldots, n]$, the number of k-combinations out of n elements is defined as the corresponding binomial coefficient:*

$$\binom{n}{k} = \frac{n!}{k!(n-k)!}. \tag{4.1}$$

Note that for $k \in [0, 1, \ldots, n]$, Eq. (4.1) characterizes the number of k-combinations out of the n elements. Throughout this chapter we will apply Eq. (4.1) to determine and describe several Pascal triangle's properties by coalescing Figure (4.2) together with Figure (4.4) that will examine the triangle's horizontal rows that lead to the Horizontally Oriented Identities. In addition, we will combine Figure (4.3) together with Figure (4.4) that will decipher the Triangle's diagonals that will direct to the Diagonally Oriented Identities. The upcoming example will portray the applications of (4.1) to calculate the triangle's elements.

Example 4.1. *Figure (4.2) decomposes Pascal's triangle into blue and red rows as shown in the corresponding sketch below:*

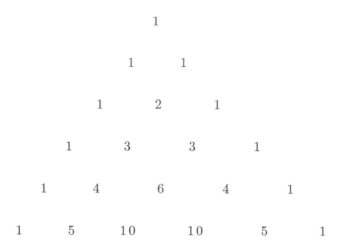

From Figure (4.4), our aim is to compute each element of Pascal's triangle by applying Eq. (4.1). For instance, by applying Eq. (4.1) we obtain all the elements of the triangle's third row:

$$\binom{3}{0} = \frac{3!}{0!3!} = 1, \quad \binom{3}{1} = \frac{3!}{1!2!} = 3, \quad \binom{3}{2} = \frac{3!}{2!1!} = 3, \quad \binom{3}{3} = \frac{3!}{3!0!} = 1.$$

The consequent example will apply Eq. (4.1) to enumerate all the combinations two out of five. Furthermore, we will coalesce (3.4) together with (4.1).

Example 4.2. *List all the two combinations out of five from the set $\{a, b, c, d, e\}$ and expresses all the combinations in terms of Eqs. (4.1) and (3.4).*

Solution: *We will structure all the possible combinations in four rows in the following configuration:*

$$\{a, b\}, \ \{a, c\}, \ \{a, d\}, \ \{a, e\}$$
$$\{b, c\}, \ \{b, d\}, \ \{b, e\}$$
$$\{c, d\}, \ \{c, e\}$$
$$\{d, e\}$$

By adding all the terms from each row we procure the following sum:

$$1 + 2 + 3 + 4 = 10 = \frac{4 \cdot 5}{2} = \binom{5}{2} = \frac{5!}{2!3!}. \qquad (4.2)$$

Therefore, for all $n \in \mathbb{N}$ we can extend Eq. (4.2) to the following result:

$$\sum_{i=1}^{n} i = \frac{n \cdot [n+1]}{2} = \binom{n+1}{2}. \qquad (4.3)$$

Eq. (4.3) will emerge as one of the triangle's properties in Example (4.7) and will direct us to supplemental Diagonally Oriented Identities. The next section will examine the triangle's Horizontally Oriented Identities.

4.1 HORIZONTALLY ORIENTED IDENTITIES

This section will focus on the triangle's **Horizontally Oriented Identities** by applying Figure (4.2) together with Figure (4.4). We will commence with the first example that renders the triangle's **Symmetry Identity**.

Example 4.3. *The following diagram depicts the **Symmetry Pattern** with the blue, green and red colors:*

$$
\begin{array}{ccccccccccccccc}
& & & & & & & 1 & & & & & & & \\
& & & & & & 1 & & 1 & & & & & & \\
& & & & & 1 & & 2 & & 1 & & & & & \\
& & & & 1 & & 3 & & 3 & & 1 & & & & \\
& & & 1 & & 4 & & 6 & & 4 & & 1 & & & \\
& & 1 & & 5 & & 10 & & 10 & & 5 & & 1 & & \\
& 1 & & 6 & & 15 & & 20 & & 15 & & 6 & & 1 & \\
1 & & 7 & & 21 & & 35 & & 35 & & 21 & & 7 & & 1 \\
\end{array}
$$

FIGURE 4.5   Pascal's triangle symmetry property.

From Figure (4.5) and via Eq. (4.1), the triangle's blue terms render the following configuration:

$$3 \ = \ 3, \quad 4 = 4, \quad 5 \ = \ 5, \quad 6 \ = \ 6, \quad 7 = 7$$

$$\binom{3}{1} = \binom{3}{2}, \quad \binom{4}{1} = \binom{4}{3}, \quad \binom{5}{1} = \binom{5}{4}, \quad \binom{6}{1} = \binom{6}{5}, \quad \binom{7}{1} = \binom{7}{6}.$$

$$(4.4)$$

Notice from Eq. (4.4) we acquire: $1 + 2 = 3$, $1 + 3 = 4$, $1 + 4 = 5$, *and* $1 + 5 = 6$. *In addition, via Eqs. (4.4) and (4.1) we obtain:*

$$\binom{3}{1} = \frac{3!}{1!2!} = \binom{3}{2}, \quad \binom{4}{1} = \frac{4!}{1!3!} = \binom{4}{3}, \quad \binom{5}{1} = \frac{5!}{4!1!} = \binom{5}{4}.$$

From Figure (4.5) and via Eq. (4.1), the triangle's green terms resemble the corresponding structure:

$$10 \ = \ 10, \quad 15 = 15, \quad 21 \ = \ 21$$

$$\binom{5}{2} = \binom{5}{3}, \quad \binom{6}{2} = \binom{6}{4}, \quad \binom{7}{2} = \binom{7}{5}.$$

$$(4.5)$$

From Eq. (4.5) we procure: $2 + 3 = 5$, $1 + 3 = 4$, $2 + 4 = 6$, *and* $2 + 5 = 7$. *Via Eqs. (4.5) and (4.1) we get:*

$$\binom{5}{2} = \frac{5!}{2!3!} = \binom{5}{3}, \quad \binom{6}{2} = \frac{6!}{2!4!} = \binom{6}{4}, \quad \binom{7}{2} = \frac{7!}{2!5!} = \binom{7}{5}, \quad \dots$$

From Figure (4.5) and via Eq. (4.1), the triangle's red terms depict the corresponding pattern:

$$35 \ = \ 35$$

$$\binom{7}{3} = \binom{7}{4},$$

$$(4.6)$$

where $3 + 4 = 7$. *Via Eqs. (4.6) and (4.1) we obtain:*

$$\binom{7}{3} = \frac{7!}{3!4!} = \binom{7}{4}.$$

Hence via Eqs. (4.4), (4.5) and (4.6), for all $n \geq 0$ *and* $k = 0, 1, \dots, n$ *we procure:*

$$\binom{n}{k} = \binom{n}{n-k},$$

$$(4.7)$$

where $k + (n - k) = n$ for all $k = 0, 1, \ldots, n$. Thus, via Eq. (4.1) we obtain:

$$\binom{n}{n-k} = \frac{n!}{(n-k)![n-(n-k)]!} = \frac{n!}{(n-k)!k!} = \binom{n}{k}.$$

Hence, the result follows.

The consequent example will depict the triangle's **Pascal's Identity** by adding two neighboring horizontal terms in each row.

Example 4.4. *The following diagram renders the **Pascal's Identity** with blue and red colors (where each red term is the sum of the two adjacent horizontal blue terms):*

```
                           1

                       1       1

                    1       2       1

                 1       3       3       1

              1       4       6       4       1

           1       5      10      10       5       1

        1       6      15      20      15       6       1

     1       7      21      35      35      21       7       1
```

FIGURE 4.6   Representation of the triangle's Pascal's identity.

When adding two adjoining horizontal blue terms in each row in Figure (4.6) together with Eq. (4.1), we procure the following pattern:

$$1 + 2 = 3, \quad 6 + 4 = 10, \quad 15 + 6 = 21, \ldots$$

$$\binom{2}{0} + \binom{2}{1} = \binom{3}{1}, \quad \binom{4}{2} + \binom{4}{3} = \binom{5}{3}, \quad \binom{6}{4} + \binom{6}{5} = \binom{7}{5}, \ldots$$

$$(4.8)$$

By applying Eq. (4.1), we acquire:

$$\binom{6}{4} + \binom{6}{5} = \frac{6!}{2!4!} + \frac{6!}{5!1!} = \frac{6!5}{2!5!} + \frac{6!2}{5!2!} = \frac{6![5+2]}{5!2!} = \frac{7!}{5!2!} = \binom{7}{5}.$$

*In addition, from Eq. (4.8), for all $n \in \mathbb{N}$ and $k \in [0, 1, \ldots, n-1]$ we obtain the corresponding **Pascal's Identity**:*

$$\binom{n}{k} + \binom{n}{k+1} = \binom{n+1}{k+1}. \tag{4.9}$$

*Proving Eq. (4.9) (**Pascal's Identity**) will be left as an end-of-chapter exercise. Eq. (4.9) will also be used to prove further identities.*

In Example (4.4) we added two horizontal neighboring terms in each row and obtained the **Pascal's Identity**. The next example will affix all the horizontal terms in each row and produce the triangle's **Power Identity** (adding up to powers of two).

Example 4.5. *In Example (4.4) we added two neighboring horizontal terms. We will decompose the triangle's rows where the blue terms render the even-ordered rows, while the red terms depict the odd-ordered rows.*

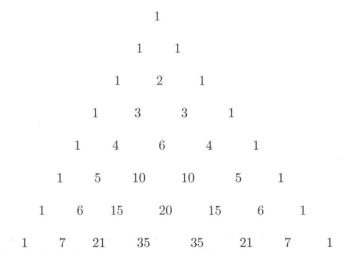

FIGURE 4.7 Representation of the triangle's power identity.

Now we will combine all the terms in each row. From Figure (4.7), by combining all the terms in each row starting with the 0th row and by applying Eq. (4.1) we obtain the following properties:

$$1 = \binom{0}{0} = 2^0 \quad \text{(0th row)},$$

$$1 + 1 = \binom{1}{0} + \binom{1}{1} = 2^1 \quad \text{(1st row)},$$

$$1 + 2 + 1 = \binom{2}{0} + \binom{2}{1} + \binom{2}{2} = 2^2 \quad \text{(2nd row)}, \qquad (4.10)$$

$$1 + 3 + 3 + 1 = \binom{3}{0} + \binom{3}{1} + \binom{3}{2} + \binom{3}{3} = 2^3 \quad \text{(3rd row)},$$

$$\vdots$$

Notice that the power of 2 corresponds directly to the order of each row. Hence, for $n \in \mathbb{N}$, (4.10) extends to the corresponding **Power Identity***:*

$$\sum_{i=0}^{n} \binom{n}{i} = 2^n. \qquad (4.11)$$

Note that the blue even-ordered rows have an odd number of terms, and the red odd-ordered rows have an even number of terms. Thus, proving Eq. (4.11) will require two cases when n is even and when n is odd. In addition, Eq. (4.11) is proved by induction by applying the **Symmetry Identity** *and* **Pascal's Identity***. Furthermore, from Figure (4.7), and via Eqs. (4.10) and (4.11), by adding all the elements of Pascal's triangle we procure:*

$$2^0 + 2^1 + 2^1 + \ldots + 2^n = \sum_{i=0}^{n} 2^i = 2^{n+1} - 1.$$

4.2 DIAGONALLY ORIENTED IDENTITIES

This section will focus on the triangle's **Diagonally Oriented Identities** by applying Figure (4.3) together with Figure (4.4). The succeeding examples will combine the adjacent diagonal terms in comparison to combining horizontal terms as we did in the previous section. In Example (4.4) we added two horizontal neighbors and obtained the **Pascals' Identity**. Analogously, the next example will initiate the **Square Identity** by adding two neighboring terms in the second diagonal.

Example 4.6. *Our aim is to derive the* **Square Identity** *by combining two neighboring blue diagonal terms in the second diagonal as shown in the diagram below:*

$$
\begin{array}{ccccccccccccccc}
& & & & & & & 1 & & & & & & & \\
& & & & & & 1 & & 1 & & & & & & \\
& & & & & 1 & & 2 & & 1 & & & & & \\
& & & & 1 & & 3 & & 3 & & 1 & & & & \\
& & & 1 & & 4 & & 6 & & 4 & & 1 & & & \\
& & 1 & & 5 & & 10 & & 10 & & 5 & & 1 & & \\
& 1 & & 6 & & 15 & & 20 & & 15 & & 6 & & 1 & \\
1 & & 7 & & 21 & & 35 & & 35 & & 21 & & 7 & & 1
\end{array}
$$

FIGURE 4.8   Triangle's second diagonal and the square identity.

Applying Figure (4.8) and implementing Eq. (4.1) to express the sum of two adjacent blue terms, we procure the following identities:

$$
1 + 3 = \binom{2}{0} + \binom{3}{1} = 2^2,
$$

$$
3 + 6 = \binom{3}{1} + \binom{4}{2} = 3^2, \tag{4.12}
$$

$$
6 + 10 = \binom{4}{2} + \binom{5}{3} = 4^2,
$$

$$
10 + 15 = \binom{5}{3} + \binom{6}{4} = 5^2,
$$

$$
\vdots
$$

Hence from Eq. (4.12), for all $n \geq 2$ we acquire the following **Square Identity:**

$$
\binom{n}{n-2} + \binom{n+1}{n-1} = n^2. \tag{4.13}
$$

Verifying Eq. (4.13) will be left as an end-of-chapter exercise.

The upcoming example will evince (3.4) as one of the properties of Pascal's triangle by summing all the consecutive diagonal terms in the first diagonal.

Example 4.7. *The corresponding scheme renders the red terms of the triangle's first diagonal and lists all the consecutive positive integers:*

$$
\begin{array}{ccccccccccccccc}
& & & & & & & 1 & & & & & & & \\
& & & & & & 1 & & 1 & & & & & & \\
& & & & & 1 & & 2 & & 1 & & & & & \\
& & & & 1 & & 3 & & 3 & & 1 & & & & \\
& & & 1 & & 4 & & 6 & & 4 & & 1 & & & \\
& & 1 & & 5 & & 10 & & 10 & & 5 & & 1 & & \\
& 1 & & 6 & & 15 & & 20 & & 15 & & 6 & & 1 & \\
1 & & 7 & & 21 & & 35 & & 35 & & 21 & & 7 & & 1 \\
\end{array}
$$

FIGURE 4.9   Triangle's first diagonal indicated in red.

In Figure (4.9), by summing all the consecutive red terms in the first diagonal and applying Eq. (4.1) we procure the following relations:

$$1+2 = \binom{1}{0} + \binom{2}{1} = \binom{3}{1} = 3,$$

$$1+2+3 = \binom{1}{0} + \binom{2}{1} + \binom{3}{2} = \binom{4}{2} = 6,$$

$$1+2+3+4 = \binom{1}{0} + \binom{2}{1} + \binom{3}{2} + \binom{4}{3} = \binom{5}{3} = 10,$$

$$1+2+3+4+5 = \binom{1}{0} + \binom{2}{1} + \binom{3}{2} + \binom{4}{3} + \binom{5}{4} = \binom{6}{4} = 15,$$

$$\vdots$$

$$(4.14)$$

From the relations in Eq. (4.14), for all n ≥ 2 we rewrite (3.4) as:

$$\binom{1}{0} + \binom{2}{1} + \binom{3}{2} + \binom{4}{3} + \cdots + \binom{n}{n-1}$$

$$= \sum_{i=0}^{n-1} \binom{i+1}{i} = \binom{n+1}{n-1}. \tag{4.15}$$

We will prove Eq. (4.15) by induction together with the **Pascal's Identity** *and will be left as an end-of-chapter exercise.*

Similar to Eq. (4.15) in Example (4.7), by combining all the blue terms in the triangle's **second diagonal** we obtain:

$$\binom{2}{0} + \binom{3}{1} + \binom{4}{2} + \binom{5}{3} + \cdots + \binom{n+1}{n-1}$$

$$= \sum_{i=0}^{n-1} \binom{i+2}{i} = \binom{n+2}{n-1}. \tag{4.16}$$

By summing all the red terms in the triangle's **third diagonal** we procure:

$$\binom{3}{0} + \binom{4}{1} + \binom{5}{2} + \binom{6}{3} + \cdots + \binom{n+2}{n-1}$$

$$= \sum_{i=0}^{n-1} \binom{i+3}{i} = \binom{n+3}{n-1}. \tag{4.17}$$

By adding all the blue terms in the triangle's **fourth diagonal** we acquire:

$$\binom{4}{0} + \binom{5}{1} + \binom{6}{2} + \binom{7}{3} + \cdots + \binom{n+3}{n-1}$$

$$= \sum_{i=0}^{n-1} \binom{i+4}{i} = \binom{n+4}{n-1}. \tag{4.18}$$

Therefore, for all $k \in \mathbb{N}$, by affixing all the terms in the kth diagonal, via Eqs. (4.15)–(4.18) we produce the corresponding identity:

$$\binom{k}{0} + \binom{k+1}{1} + \binom{k+2}{2} + \cdots + \binom{n+k-1}{n-1}$$

$$= \sum_{i=0}^{n-1} \binom{i+k}{i} = \binom{n+k}{n-1}. \tag{4.19}$$

Proving Eq. (4.19) by induction will be left as an end-of-chapter exercise.

4.3 CHAPTER 4 EXERCISES

In problems 1–10, **simplify** the following expressions:

1: $\dfrac{9!}{3!6!}$

2: $\dfrac{10!}{5!5!}$

3: $\dbinom{6}{4}\dbinom{4}{2}$

4: $\dbinom{8}{2}\dbinom{6}{4}$

5: $\dfrac{7}{3}\dbinom{6}{2}$

6: $\dfrac{5}{8}\dbinom{8}{3}$

7: $\dfrac{(k+2)!}{k!}, \; k \in \mathbb{N}$

8: $\dfrac{(k+3)!}{k!}, \; k \in \mathbb{N}$

9: $\dfrac{(k+n)!}{k!}, \; k, n \in \mathbb{N}$

10: $k \cdot [k! + (k-1)!], \; k \in \mathbb{N}$

11: $\dfrac{8!}{4!}$

12: $\dfrac{12!}{6!}$

13: $\dfrac{(2k)!}{k!}, \; k \in \mathbb{N}$

14: $\dfrac{\dbinom{n}{k+1}}{\dbinom{n}{k}}, \; k, n \in \mathbb{N}$

In problems 15–22, using Eq. (4.1) **prove** the following expressions:

15: $\dbinom{2k}{k}$ is even, $k \in \mathbb{N}$

16: $\dbinom{2k}{2} = 2\dbinom{k}{2} + k^2$, $k \in \mathbb{N}$

17: $\dbinom{3k}{3} = 3\dbinom{k}{3} + 6k\dbinom{k}{2} + k^3$, $k \in \mathbb{N}$

18: $\dbinom{n}{k} = \dfrac{n}{k}\dbinom{n-1}{k-1}$, $k, n \in \mathbb{N}$

19: $\dbinom{n}{k} = \dfrac{n}{n-k}\dbinom{n-1}{k}$, $k, n \in \mathbb{N}$

20: $\dbinom{n}{k} + \dbinom{n}{k+1} = \dbinom{n+1}{k+1}$, $k, n \in \mathbb{N}$

21: $\dbinom{n}{k}\dbinom{k}{j} = \dbinom{n}{j}\dbinom{n-j}{k-j}$, $j \le k \le n$

22: $\dbinom{n}{n-2} + \dbinom{n+1}{n-1} = n^2$, $n \ge 2$

In problems 23–30, prove the following expressions by **induction**:

23: $\displaystyle\sum_{i=0}^{n-1} \dbinom{i+1}{i} = \dbinom{n+1}{n-1}$, $n \ge 2$

24: $\displaystyle\sum_{i=0}^{n-1} \dbinom{i+2}{i} = \dbinom{n+2}{n-1}$, $n \ge 2$

25: $\displaystyle\sum_{i=0}^{n-1} \dbinom{i+3}{i} = \dbinom{n+3}{n-1}$, $n \ge 2$

26: $\displaystyle\sum_{i=0}^{n-1} \dbinom{i+4}{i} = \dbinom{n+4}{n-1}$, $n \ge 2$

27: $\displaystyle\sum_{i=0}^{n-1} \binom{i+k}{i} = \binom{n+k}{n-1}$, $n \geq 2, k \in \mathbb{N}$

28: $\displaystyle\sum_{i=0}^{n} \binom{n}{i} \overset{\cdot}{=} 2^n$, $n \in \mathbb{N}$

29: Using Problem 28, prove:

$$\sum_{i=0}^{n} \binom{n}{i} 2^i = 3^n, \ n \in \mathbb{N}$$

30: Using Problem 28, prove:

$$\sum_{i=0}^{n} [i+1] \binom{n}{i} = 2^n + n2^{n-1}, \ n \in \mathbb{N}$$

31: Using Problem 28, prove:

$$\sum_{i=0}^{n} \binom{2n+1}{i} = 2^{2n}, \ n \in \mathbb{N}$$

CHAPTER 5

First-Order Recursive Relations

THIS CHAPTER'S AIMS ARE to characterize sequences encountered in the previous chapters as recursive relations while introducing the first-order recursive sequences. Recall various sorts of sequences that we examined from the proceeding chapters:

1. Linear Sequence: 1, 4, 7, 10, 13, ...

2. Summation-Type Sequence: 3, 6, 10, 15, 21, ...

3. Geometric Sequence: 5, 20, 80, 320, 1280, ...

4. Factorial-Type Sequence: 3, 6, 24, 144, 1152, ...

Our intents are to alternatively describe these sequences and analogous sequences as **recursive relations**. This will then direct us to deeper understanding of the new categories of sequences and their unique traits. We will emerge our study of a linear sequence expressed as a **recursive formula** and as an **Initial Value Problem**.

Example 5.1. *Write a recursive formula for:*

$$3, \ 7, \ 11, \ 15, \ 19, \ 23, \ 27, \dots . \tag{5.1}$$

Solution: *In Eq. (5.1), we shift from neighbor to neighbor by adding a four. We can write the following formula of (5.1):*

$$\{4n + 3\}_{n=0}^{\infty}.$$

We will alternatively express (5.1) as a recursive sequence. By iteration and induction we acquire:

$$x_0 = 3,$$
$$x_0 + 4 = 3 + 4 = 7 = x_1,$$
$$x_1 + 4 = 7 + 4 = 11 = x_2,$$
$$x_2 + 4 = 11 + 4 = 15 = x_3,$$
$$x_3 + 4 = 15 + 4 = 19 = x_4,$$
$$x_4 + 4 = 19 + 4 = 23 = x_5,$$
$$\vdots$$

For all $n \geq 0$ we obtain the following **Initial Value Problem** *rendering (5.1):*

$$\begin{cases} x_{n+1} = x_n + 4, \\ x_0 = 3. \end{cases}$$

This is a special case of a First-Order Linear Nonhomogeneous Difference Equation.

The successive two examples will describe a **Summation-Type** Sequence recursively.

Example 5.2. *Write a recursive formula for:*

$$1, 3, 6, 10, 15, 21, 28, \ldots. \tag{5.2}$$

Solution: *Eq. (5.2) depicts the sum of the consecutive positive integers starting at one traced by the corresponding formula:*

$$\sum_{i=1}^{n} i.$$

We will formulate (5.2) as a recursive sequence. By iteration we obtain:

$$x_0 = 1,$$
$$x_0 + 2 = 1 + 2 = 3 = x_1,$$
$$x_1 + 3 = 3 + 3 = 6 = x_2,$$
$$x_2 + 4 = 6 + 4 = 10 = x_3,$$
$$x_3 + 5 = 10 + 5 = 15 = x_4,$$
$$x_4 + 6 = 15 + 6 = 21 = x_5,$$
$$\vdots$$

For all $n \geq 0$ we acquire the related **Initial Value Problem** *describing (5.2):*

$$\begin{cases} x_{n+1} = x_n + (n + 2), \\ \quad x_0 = 1. \end{cases}$$

This is a special case of a First-Order Linear Nonautonomous Difference Equation in the **additive form**.

Example 5.3. *Write a recursive formula for:*

$$1, \ 4, \ 9, \ 16, \ 25, \ 36, \ 49, \ldots. \tag{5.3}$$

Solution: *Eq. (5.3) renders perfect squares starting at one characterized by the following formula:*

$$\{n^2\}_{n=1}^{\infty}.$$

Now we will formulate (5.3) as a recursive sequence. By iteration we obtain:

$$x_0 = 1,$$
$$x_0 + 3 = 1 + 3 = 4 = x_1,$$
$$x_1 + 5 = 4 + 5 = 9 = x_2,$$
$$x_2 + 7 = 9 + 7 = 16 = x_3,$$
$$x_3 + 9 = 16 + 9 = 25 = x_4,$$
$$x_4 + 11 = 25 + 11 = 36 = x_5,$$
$$\vdots$$

For all $n \geq 0$ we procure the ensuing **Initial Value Problem** *depicting (5.3):*

$$\begin{cases} x_{n+1} = x_n + (2n + 3), \\ \quad x_0 = 1. \end{cases}$$

This is another special case of a First-Order Linear Nonautonomous Difference Equation in the **additive form**.

The consequent example will describe a **Geometric Sequence** recursively.

Example 5.4. *Write a recursive formula for:*

$$11, \ 22, \ 44, \ 88, \ 176, \ 352, \ 704, \ldots. \tag{5.4}$$

Solution: *In Eq. (5.4) we shift from neighbor to neighbor by multiplying by two. Hence, we can write the following formula of (5.4):*

$$\{11 \cdot 2^n\}_{n=0}^{\infty}.$$

Furthermore, we will treat (5.4) as a recursive sequence. Notice:

$$x_0 = 11,$$

$$x_0 \cdot 2 = 11 \cdot 2 = 22 = x_1,$$

$$x_1 \cdot 2 = 22 \cdot 2 = 44 = x_2,$$

$$x_2 \cdot 2 = 44 \cdot 2 = 88 = x_3,$$

$$x_3 \cdot 2 = 88 \cdot 2 = 176 = x_4,$$

$$x_4 \cdot 2 = 176 \cdot 2 = 352 = x_5,$$

$$\vdots$$

Thus, for all $n \geq 0$ we obtain the corresponding **Initial Value Problem** *tracing (5.4):*

$$\begin{cases} x_{n+1} = 2x_n, \\ \quad x_0 = 11. \end{cases}$$

This is a special case of a First-Order Linear Homogeneous Difference Equation.

The sequential example will describe the **Factorial Pattern** recursively.

Example 5.5. *Write a recursive formula for:*

$$1, \ 2, \ 6, \ 24, \ 120, \ 720, \ 5040, \dots \quad (5.5)$$

Solution: *We render Eq. (5.5) with the corresponding iterative pattern:*

$$x_0 = 1,$$
$$x_0 \cdot 2 = 1 \cdot 2 = 2 = x_1,$$
$$x_1 \cdot 3 = 2 \cdot 3 = 6 = x_2,$$
$$x_2 \cdot 4 = 6 \cdot 4 = 24 = x_3,$$
$$x_3 \cdot 5 = 24 \cdot 5 = 120 = x_4,$$
$$x_4 \cdot 6 = 120 \cdot 6 = 720 = x_5,$$
$$\vdots$$

Thus, for all $n \geq 0$ we obtain the successive **Initial Value Problem** *describing (5.5):*

$$\begin{cases} x_{n+1} = (n+2) \cdot x_n, \\ \quad x_0 = 1. \end{cases}$$

This example describes the **Factorial Pattern** *and is a special case of a First-Order Linear Nonautonomous Difference Equation in the* **multiplicative form**.

Now we will transition to solving recursive sequences explicitly by inductively obtaining a formula and solving an **Initial Value Problem**.

5.1 FIRST-ORDER LINEAR RECURSIVE SEQUENCES

This section's aim is to inductively determine an explicit solution to assorted recursive sequences. We will examine varieties of first-order recursive sequences such as:

1. Linear Homogeneous: $x_{n+1} = 4x_n, \quad n = 0, 1, \dots$.

2. Linear Nonhomogeneous: $x_{n+1} = 3x_n + 6, \quad n = 0, 1, \dots$.

3. Linear Nonautonomous: $x_{n+1} = a_n x_n, \quad n = 0, 1, \dots$.

4. Linear Nonautonomous: $x_{n+1} = x_n + b_n, \quad n = 0, 1, \dots$.

We will commence with obtaining an explicit solution to a linear homogeneous recursive relation whose solution depicts a geometric sequence:

$$x_{n+1} = ax_n, \quad n = 0, 1, \ldots, \tag{5.6}$$

where $a \neq 0$. By iterations and induction we acquire the following pattern:

$$
\begin{aligned}
&x_0, \\
&x_1 = ax_0, \\
&x_2 = ax_1 = a \cdot [ax_0] = a^2 x_0, \\
&x_3 = ax_2 = a \cdot [a^2 x_0] = a^3 x_0, \\
&x_4 = ax_3 = a \cdot [a^3 x_0] = a^4 x_0, \\
&x_5 = ax_4 = a \cdot [a^4 x_0] = a^5 x_0, \\
&\vdots
\end{aligned}
$$

Hence, for all $n \in \mathbb{N}$ we procure the corresponding solution of Eq. (5.6):

$$x_n = a^n x_0. \tag{5.7}$$

Equation (5.7) is acquired by induction to a First-Order Linear Homogeneous Difference Equation. The consequent example will illustrate the application of Eq. (5.6) in solving an **Initial Value Problem.**

Example 5.6. *Solve the given* **Initial Value Problem***:*

$$
\begin{cases}
3x_{n+1} - 2x_n = 0, & n = 0, 1, \ldots, \\
x_0 = \frac{8}{27},
\end{cases}
$$

and verify that the solution is correct.

Solution: *From Eq. (5.7) we obtain the following solution:*

$$x_n = x_0 \left(\frac{2}{3}\right)^n = \frac{8}{27} \left(\frac{2}{3}\right)^n = \left(\frac{2}{3}\right)^{n+3}. \tag{5.8}$$

Now via (5.8) we acquire:

$$3x_{n+1} - 2x_n = 3\left(\frac{2}{3}\right)^{n+4} - 2\left(\frac{2}{3}\right)^{n+3} = \frac{2^{n+4}}{3^{n+3}} - \frac{2^{n+4}}{3^{n+3}} = 0.$$

Hence, the result follows.

We will next proceed with obtaining an explicit solution to a linear nonhomogeneous recursive relation in the form:

$$x_{n+1} = ax_n + b, \quad n = 0, 1, \ldots, \tag{5.9}$$

where $a \neq 0, 1$ and $b \neq 0$. By iterations and induction we procure:

$$x_0,$$

$$x_1 = ax_0 + b,$$

$$x_2 = ax_1 + b = a \cdot [ax_0 + b] = a^2 x_0 + ab,$$

$$x_3 = ax_2 + b = a \cdot [a^2 x_0 + b] = a^3 x_0 + a^2 b + ab + b,$$

$$x_4 = ax_3 + b = a \cdot [a^3 x_0 + b] = a^4 x_0 + a^3 b + a^2 b + ab + b,$$

$$x_5 = ax_4 + b = a \cdot [a^2 x_0 + b] = a^5 x_0 + a^4 b + a^3 b + a^2 b + ab + b,$$

$$\vdots$$

Thus, for all $n \in \mathbb{N}$ we acquire the following solution of Eq. (5.9):

$$x_n = a^n x_0 + b \left[\sum_{i=0}^{n-1} a^i \right] = a^n x_0 + b \left[\frac{1 - a^n}{1 - a} \right]. \tag{5.10}$$

Equation (5.10) is obtained by induction to a First-Order Linear Non-homogeneous Difference Equation. We can also reformulate (5.10) as:

$$x_n = a^n \left[x_0 - \frac{b}{1 - a} \right] + \frac{b}{1 - a}. \tag{5.11}$$

The consequent example will illustrate the application of Eq. (5.11) in solving an **Initial Value Problem** of Eq. (5.9).

Example 5.7. *Solve the designated* **Initial Value Problem**:

$$\begin{cases} 4x_{n+1} - x_n = 9, & n = 0, 1, \ldots, \\ x_0 = 7, \end{cases}$$

and verify that the solution is correct.

Solution: *From Eq. (5.11) we obtain:*

$$x_n = \left(\frac{1}{4} \right)^n [7 - 3] + 3 = \left(\frac{1}{4} \right)^{n-1} + 3. \tag{5.12}$$

From Eq. (5.12) we get:

$$4x_{n+1} - x_n = 4\left[\left(\frac{1}{4}\right)^n + 3\right] - \left[\left(\frac{1}{4}\right)^{n-1} + 3\right]$$
$$= \frac{1}{4^{n-1}} + 12 - \frac{1}{4^{n-1}} - 3 = 9.$$

Thus, the result follows.

The upcoming sequence of examples will focus on acquiring a general solution to supplemental categories of recursive relations.

Example 5.8. *Determine an* **explicit solution** *to the following recursive relation:*

$$x_{n+1} = x_n + b, \quad n = 0, 1, \ldots,$$

where $b \neq 0$.

Solution: *By iterations we acquire the following pattern:*

$$x_0,$$
$$x_1 = x_0 + b,$$
$$x_2 = x_1 + b = [x_0 + b] + b = x_0 + 2b,$$
$$x_3 = x_2 + b = [x_0 + 2b] + b = x_0 + 3b,$$
$$x_4 = x_3 + b = [x_0 + 3b] + b = x_0 + 4b,$$
$$x_5 = x_4 + b = [x_0 + 4b] + b = x_0 + 5b,$$
$$\vdots$$

Hence, for all $n \in \mathbb{N}$:

$$x_n = x_0 + nb. \tag{5.13}$$

Equation (5.13) is determined by induction to the First-Order Linear Non-homogeneous Recursive Sequence. Verifying Eq. (5.13) will be left as an end-of-chapter exercise.

Example 5.9. *Determine an* **explicit solution** *to the following recursive relation:*

$$x_{n+1} = -x_n + b, \quad n = 0, 1, \ldots,$$

where $b \neq 0$.

Solution: *By iterations we acquire the following pattern:*

$$x_0,$$
$$x_1 = -x_0 + b,$$
$$x_2 = -x_1 + b = -[-x_0 + b] + b = x_0,$$
$$x_3 = -x_2 + b = -[x_0] + b = -x_0 + b,$$
$$x_4 = -x_3 + b = -[-x_0 + b] + b = x_0,$$
$$\vdots$$

Hence, for all $n \geq 0$:

$$x_n = \begin{cases} x_0 & \text{if } n \text{ is even,} \\ -x_0 + b & \text{if } n \text{ is odd.} \end{cases}$$

This is an example of a period-2 cycle and a piecewise solution to a First-Order Linear Nonhomogeneous Recursive Sequence.

We will examine more examples of periodic recursive relations that exhibit periodic cycles in Chapter 6. The next section will focus on obtaining an explicit solution to a First-Order Linear Nonautonomous Difference Equation.

5.2 NONAUTONOMOUS RECURSIVE SEQUENCES

This section's aim is to inductively determine an explicit solution to specific nonautonomous recursive sequences. We will examine varieties of first-order nonautonomous linear recursive sequences as:

1. Linear Nonautonomous: $x_{n+1} = x_n + b_n$, $n = 0, 1, \ldots,$

2. Linear Nonautonomous: $x_{n+1} = a_n x_n$, $n = 0, 1, \ldots,$

3. Linear Nonautonomous: $x_{n+1} = a_n x_n + b_n$, $n = 0, 1, \ldots,$

where $\{a_n\}_{n=0}^{\infty}$ and $\{b_n\}_{n=0}^{\infty}$ are sequences of real numbers. In (1), for all $n \geq 0$ each term of the sequence $\{b_n\}_{n=0}^{\infty}$ is added to x_n during each iteration. In (2), on the other hand, for all $n \geq 0$ each term of the sequence $\{a_n\}_{n=0}^{\infty}$ is multiplied by x_n during each iteration. In (3), for all $n \geq 0$, b_n is added to the product $a_n x_n$.

First we will commence with the nonautonomous homogeneous linear recursive sequence in the **additive form**:

$$x_{n+1} = x_n + b_n, \quad n = 0, 1, \ldots, \tag{5.14}$$

where $\{b_n\}_{n=0}^{\infty}$ is a sequence of real numbers. By iterations we acquire:

$$x_0,$$

$$x_1 = x_0 + b_0,$$

$$x_2 = x_1 + b_1 = x_0 + [b_0 + b_1],$$

$$x_3 = x_2 + b_2 = x_0 + [b_0 + b_1 + b_2],$$

$$x_4 = x_3 + b_3 = x_0 + [b_0 + b_1 + b_2 + b_3],$$

$$x_5 = x_4 + b_4 = x_0 + [b_0 + b_1 + b_2 + b_3 + b_4],$$

$$\vdots$$

For all $n \in \mathbb{N}$, we obtain the corresponding solution of Eq. (5.14):

$$x_n = x_0 + [b_0 + b_1 + \ldots + b_{n-1}] = x_0 + \sum_{i=0}^{n-1} b_i. \tag{5.15}$$

In the upcoming two examples, we will work out the details in solving the nonautonomous nonhomogeneous first-order recursive relations.

Example 5.10. *Determine an **explicit solution** of the following recursive relation:*

$$x_{n+1} = x_n + (n + 1), \quad n = 0, 1, \ldots.$$

Solution: *By iterations we obtain:*

$$x_0,$$
$$x_1 = x_0 + 1,$$
$$x_2 = x_1 + 2 = x_0 + [1 + 2],$$
$$x_3 = x_2 + 3 = x_0 + [1 + 2 + 3],$$
$$x_4 = x_3 + 4 = x_0 + [1 + 2 + 3 + 4],$$
$$x_5 = x_4 + 5 = x_0 + [1 + 2 + 3 + 4 + 5],$$
$$\vdots$$

From Eq. (5.15) and Eq. (3.4) for all $n \in \mathbb{N}$ we acquire:

$$x_n = x_0 + \sum_{i=1}^{n} i = x_0 + \left[\frac{n[n+1]}{2} \right].$$

Example 5.11. *Determine an* **explicit solution** *of the following recursive relation:*

$$x_{n+1} = x_n + 2^n, \quad n = 0, 1, \ldots.$$

Solution: *By iterations we acquire:*

$$
\begin{aligned}
& x_0, \\
& x_1 = x_0 + 2^0, \\
& x_2 = x_1 + 2^1 = x_0 + \left[2^0 + 2^1\right], \\
& x_3 = x_2 + 2^2 = x_0 + \left[2^0 + 2^1 + 2^2\right], \\
& x_4 = x_3 + 2^3 = x_0 + \left[2^0 + 2^1 + 2^2 + 2^3\right], \\
& x_5 = x_4 + 2^4 = x_0 + \left[2^0 + 2^1 + 2^2 + 2^3 + 2^4\right], \\
& \vdots
\end{aligned}
$$

From Eq. (5.15) and Eq. (3.4) for all $n \in \mathbb{N}$ we obtain:

$$x_n = x_0 + \sum_{i=0}^{n-1} 2^i = x_0 + \left[2^n - 1\right].$$

The consequent two examples will illustrate solutions to the recursive relations with alternating terms.

Example 5.12. *Determine an* **explicit solution** *of the following recursive relation:*

$$x_{n+1} = -x_n + 2^n, \quad n = 0, 1, \ldots.$$

Solution: *By iterations we obtain:*

$$
\begin{aligned}
& x_0, \\
& x_1 = -x_0 + 1, \\
& x_2 = -x_1 + 2 = x_0 - \left[1 - 2\right], \\
& x_3 = -x_2 + 2^2 = -x_0 + \left[1 - 2 + 2^2\right], \\
& x_4 = -x_2 + 2^3 = x_0 - \left[1 - 2 + 2^2 - 2^3\right], \\
& x_5 = -x_2 + 2^4 = -x_0 + \left[1 - 2 + 2^2 - 2^3 + 2^4\right], \\
& \vdots
\end{aligned}
$$

Hence, for all $n \in \mathbb{N}$:

$$
\begin{aligned}
x_n &= (-1)^n x_0 + (-1)^{n+1} \sum_{i=0}^{n-1} (-1)^i 2^i \\
&= (-1)^n x_0 + (-1)^{n+1} \left[\frac{1 - (-2)^n}{3}\right].
\end{aligned}
$$

Example 5.13. *Determine an* **explicit solution** *of the following recursive relation:*

$$x_{n+1} = -x_n + (n+1), \quad n = 0, 1, \ldots.$$

Solution: *By iterations we obtain:*

$$x_0,$$
$$x_1 = -x_0 + 1,$$
$$x_2 = -x_1 + 2 = x_0 - [1 - 2],$$
$$x_3 = -x_2 + 3 = -x_0 + [1 - 2 + 3],$$
$$x_4 = -x_2 + 4 = x_0 - [1 - 2 + 3 - 4],$$
$$x_5 = -x_4 + 5 = -x_0 + [1 - 2 + 3 - 4 + 5],$$
$$x_6 = -x_5 + 6 = x_0 - [1 - 2 + 3 - 4 + 5 - 6],$$
$$x_7 = -x_6 + 7 = -x_0 + [1 - 2 + 3 - 4 + 5 - 6 + 7],$$
$$\vdots$$

Hence, for all $n \in \mathbb{N}$:

$$x_n = \begin{cases} x_0 - \frac{n}{2} & \text{if } n \text{ is even,} \\ -x_0 + \frac{n+1}{2} & \text{if } n \text{ is odd.} \end{cases}$$

Next we will analyze the nonautonomous homogeneous linear recursive sequence in the **multiplicative form**:

$$x_{n+1} = a_n x_n, \quad n = 0, 1, \ldots, \tag{5.16}$$

where $\{a_n\}_{n=0}^{\infty}$ is a sequence of real numbers. By iterations we procure:

$$x_0,$$
$$x_1 = a_0 x_0,$$
$$x_2 = a_1 x_1 = [a_0 a_1] x_0,$$
$$x_3 = a_2 x_2 = [a_0 a_1 a_2] x_0,$$
$$x_4 = a_3 x_3 = [a_0 a_1 a_2 a_3] x_0,$$
$$\vdots$$

For all $n \in \mathbb{N}$, we obtain the corresponding solution of Eq. (5.16):

$$x_n = [a_0 \cdot a_1 \cdot \ldots \cdot a_{n-1}] x_0 = \left[\prod_{i=0}^{n-1} a_i \right] x_0. \tag{5.17}$$

Example 5.14. *Determine the* **explicit solution** *of the following recursive relation:*

$$x_{n+1} = 2(n+1)x_n, \quad n = 0, 1, \ldots,$$

where $x_0 \neq 0$.

Solution: *By iteration we obtain:*

$$x_0,$$
$$x_1 = 2 \cdot 1 x_0,$$
$$x_2 = 2 \cdot 2 x_1 = 2^2 \cdot [2 \cdot 1] x_0,$$
$$x_3 = 2 \cdot 3 x_2 = 2^3 \cdot [3 \cdot 2 \cdot 1] x_0,$$
$$x_4 = 2 \cdot 4 x_3 = 2^4 \cdot [4 \cdot 3 \cdot 2 \cdot 1] x_0,$$
$$x_5 = 2 \cdot 5 x_4 = 2^5 \cdot [5 \cdot 4 \cdot 3 \cdot 2 \cdot 1] x_0,$$
$$\vdots$$

Thus, for all $n \in \mathbb{N}$:

$$x_n = 2^n \left[\prod_{i-1}^{n} i \right] x_0 = 2^n n! \, x_0.$$

Example 5.15. *Determine the* **explicit solution** *of the following recursive relation:*

$$x_{n+1} = 2^{n+1} x_n, \quad n = 0, 1, \ldots,$$

where $x_0 \neq 0$.

Solution: *By iteration we acquire:*

$$x_0,$$
$$x_1 = 2 x_0,$$
$$x_2 = 2^2 x_1 = 2^{[1+2]} x_0,$$
$$x_3 = 2^3 x_2 = 2^{[1+2+3]} x_0,$$
$$x_4 = 2^4 x_3 = 2^{[1+2+3+4]} x_0,$$
$$x_5 = 2^5 x_4 = 2^{[1+2+3+4+5]} x_0,$$
$$\vdots$$

Thus, for all $n \in \mathbb{N}$:

$$x_n = \left[\prod_{i=1}^{n} 2^i\right] x_0 = 2^{\left[\sum_{i=1}^{n} i\right]} x_0 = 2^{\left[\frac{n(n+1)}{2}\right]} x_0.$$

5.3 CHAPTER 5 EXERCISES

In problems 1–8, write a **recursive formula** (as an initial value problem) of each sequence:

1:    2, 7, 12, 17, 22, 27, 32, ...

2:    7, 19, 31, 43, 55, 67, 79, ...

3:    5, 7, 11, 17, 25, 35, 47, ...

4:    4, 7, 13, 22, 34, 49, 67, ...

5:    3, 7, 15, 27, 43, 63, 87, ...

6:    2, 3, 6, 11, 18, 27, 38, ...

7:    5, 8, 17, 32, 53, 80, 113, ...

8:    1, 2, 7, 16, 29, 46, 67, ...

In problems 9–18, write a **recursive formula** (as an initial value problem) of each sequence:

9:    4, 12, 36, 108, 324, 972, ...

10:    9, 18, 36, 72, 144, 288, ...

11:    54, 36, 24, 16, $\frac{32}{3}$, $\frac{64}{9}$, ...

12:    32, 24, 18, $\frac{27}{2}$, $\frac{81}{8}$, $\frac{343}{32}$, ...

13:    1, 2, 8, 48, 384, 3840,

14:    1, 3, 15, 105, 945, 10,395, ...

15:    1, 5, 45, 585, 9945, ...

16:    $1 \cdot 2$, $2 \cdot 3$, $3 \cdot 4$, $4 \cdot 5$, $5 \cdot 6$, ...

17:    $1 \cdot 3$, $3 \cdot 5$, $5 \cdot 7$, $7 \cdot 9$, $9 \cdot 11$, ...

18:    $2 \cdot 4$, $4 \cdot 6$, $6 \cdot 8$, $8 \cdot 10$, $10 \cdot 12$, ...

In problems 19–22, write a **recursive formula** (as an initial value problem) of each summation:

19: $\sum_{k=1}^{n} (2k - 1)$.

20: $\sum_{k=0}^{n} (3k + 2)$.

21: $\sum_{k=1}^{n} k^2$.

22: $\sum_{k=0}^{n} \left(\frac{1}{2}\right)^k$.

In problems 23–28, show that the solution **satisfies** the given recursive sequence:

23: $x_n = \left(\frac{2}{5}\right)^{n+1}$ is a solution of $5x_{n+1} - 2x_n = 0$.

24: $x_n = \frac{9^{n+1}}{4^{n-1}}$ is a solution of $4x_{n+1} - 9x_n = 0$.

25: $x_n = \left(\frac{1}{3}\right)^n + 3$ is a solution of $3x_{n+1} - x_n = 6$.

26: $x_n = 5^{n+1} - 4$ is a solution of $x_{n+1} - 5x_n = 16$.

27: $x_n = (-4)^{n-2} + 3$ is a solution of $x_{n+1} + 4x_n = 15$.

28: $x_n = nb + 3$ is a solution of $x_{n+1} - x_n = b$.

In problems 29–44, solve the given **Initial Value Problem** and check your answer.

29:
$$\begin{cases} 2x_{n+1} - x_n = 0, & n = 0, 1, \ldots. \\ x_0 = \frac{1}{8}. \end{cases}$$

30:
$$\begin{cases} 3x_{n+1} - 2x_n = 0, & n = 0, 1, \ldots. \\ x_0 = \frac{4}{9}. \end{cases}$$

31:
$$\begin{cases} 3x_{n+1} - 4x_n = 0, & n = 0, 1, \ldots. \\ x_0 = \frac{16}{9}. \end{cases}$$

32:
$$\begin{cases} 5x_{n+1} - 2x_n = 6, & n = 0, 1, \ldots. \\ x_0 = 3. \end{cases}$$

33:
$$\begin{cases} 7x_{n+1} - 3x_n = 12, & n = 0, 1, \dots \\ x_0 = 4. \end{cases}$$

34:
$$\begin{cases} x_{n+1} - x_n = 4, & n = 0, 1, \dots \\ x_0 = -2. \end{cases}$$

35:
$$\begin{cases} 2x_{n+1} - 2x_n = -1, & n = 0, 1, \dots \\ x_0 = \frac{3}{2}. \end{cases}$$

36:
$$\begin{cases} 2x_{n+1} + 2x_n = 7, & n = 0, 1, \dots \\ x_0 = \frac{1}{2}. \end{cases}$$

37:
$$\begin{cases} x_{n+1} + x_n = 7, & n = 0, 1, \dots \\ x_0 = -2. \end{cases}$$

38:
$$\begin{cases} x_{n+1} - x_n = (-1)^n, & n = 0, 1, \dots \\ x_0 = 4. \end{cases}$$

39:
$$\begin{cases} x_{n+1} + x_n = (-1)^{n+1}, & n = 0, 1, \dots \\ x_0 = 4. \end{cases}$$

40:
$$\begin{cases} x_{n+1} - x_n = 3^{n+2}, & n = 0, 1, \dots \\ x_0 = 3. \end{cases}$$

41:
$$\begin{cases} x_{n+1} - x_n = 2^{n+3}, & n = 0, 1, \dots \\ x_0 = 4. \end{cases}$$

42:
$$\begin{cases} x_{n+1} - x_n = (2n + 3), & n = 0, 1, \dots \\ x_0 = 1. \end{cases}$$

43:
$$\begin{cases} x_{n+1} - x_n = (n + 2)^2, & n = 0, 1, \dots \\ x_0 = 1. \end{cases}$$

44:
$$\begin{cases} x_{n+1} - x_n = 2n + 2, & n = 0, 1, \dots \\ x_0 = 0. \end{cases}$$

CHAPTER 6

Periodic Traits

THIS CHAPTER'S PRIMARY GOAL is to investigate the periodic traits of the Autonomous and Nonautonomous First-Order Linear Recursive Relations. In some instances we will encounter the existence of unique periodic cycles. We will examine the first-order autonomous and nonautonomous linear recursive sequences that exhibit periodic behavior with different periodic cycles:

1. Linear Autonomous: $x_{n+1} = -x_n, \quad n = 0, 1, \ldots,$

2. Linear Autonomous: $x_{n+1} = -x_n + b, \quad n = 0, 1, \ldots,$

3. Linear Nonautonomous: $x_{n+1} = \pm a_n x_n, \quad n = 0, 1, \ldots,$

4. Linear Nonautonomous: $x_{n+1} = \pm x_n + b_n, \quad n = 0, 1, \ldots,$

5. Linear Nonautonomous: $x_{n+1} = \pm a_n x_n + b_n, \quad n = 0, 1, \ldots,$

where $\{a_n\}_{n=0}^{\infty}$ and $\{b_n\}_{n=0}^{\infty}$ are periodic sequences.

Definition 10. *The sequence $\{x_n\}_{n=0}^{\infty}$ is periodic with **period-p**, ($p \geq 2$), provided that for all $n \geq 0$:*

$$x_{n+p} = x_n.$$

The smallest such number p is the **minimal period**. We will commence with graphical examples of assorted periodic cycles. The corresponding sketch traces an alternating period-2 cycle:

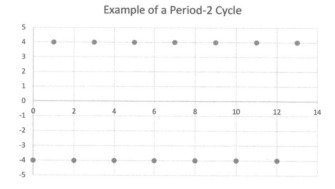

FIGURE 6.1   An alternating period-2 cycle between −4 and 4.

Figure (6.1) portrays an alternating period-2 cycle. For all $n \geq 0$, $x_{n+2} = x_n$, $x_n = -x_{n+1}$ and:

$$x_0 = x_2 = x_4 = \ldots \quad \text{and} \quad x_1 = x_3 = x_5 = \ldots$$

Figure (6.1) is also a special case of Eq. (5.6) when $a = -1$ and $x_0 = -4$. The next diagram describes an **increasing** positive period-3 cycle:

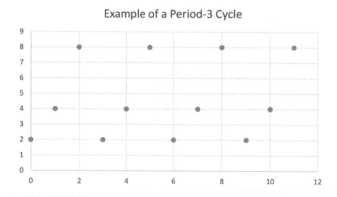

FIGURE 6.2   An increasing positive period-3 cycle with values 2, 4 and 8.

Figure (6.2) resembles an increasing period-3 cycle by a geometric pattern. For all $n \geq 0$, $x_{n+3} = x_n$. The consequent sketch describes an alternating period-6 cycle (Figure 6.3):

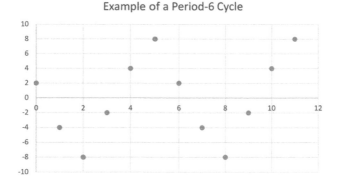

FIGURE 6.3   An alternating period-6 cycle.

Note that for all $n \geq 0$, $x_{n+6} = x_n$ and $x_n = -x_{n+3}$.

6.1 AUTONOMOUS RECURSIVE SEQUENCES

We will commence our study of periodic properties of the following first-order autonomous recursive relation:

$$x_{n+1} = -x_n + b, \quad n = 0, 1, \ldots, \tag{6.1}$$

where $x_0, b \in \Re$. Eq. (6.1) will consists of two cases when $b = 0$ and when $b \neq 0$. In fact when $b = 0$, Eq. (6.1) reduces to the corresponding homogeneous linear recursive sequence:

$$x_{n+1} = -x_n, \quad n = 0, 1, \ldots, \tag{6.2}$$

where $x_0 \neq 0$. By iterations and induction we obtain the following solution to Eq. (6.2):

$$x_n = (-1)^n x_0 = \begin{cases} x_0 & \text{if } n \text{ is even,} \\ -x_0 & \text{if } n \text{ is odd.} \end{cases} \tag{6.3}$$

Notice that Figure (1.15) in Chapter 1 is a special case of (6.3) when $x_0 = 1$ and Figure (6.1) is a special case of (6.3) when $x_0 = -4$.

Now we will assume that $b \neq 0$. Then the general solution of Eq. (6.1) is:

$$x_n = \begin{cases} x_0 & \text{if } n \text{ is even,} \\ -x_0 + b & \text{if } n \text{ is odd.} \end{cases} \tag{6.4}$$

We will now transition to periodic traits of the nonautonomous recursive relations.

6.2 NONAUTONOMOUS RECURSIVE SEQUENCES

We will proceed with the study of periodic traits of the first-order nonautonomous recursive relation in the form:

$$x_{n+1} = a_n x_n + b_n, \quad n = 0, 1, \ldots, \tag{6.5}$$

where $\{a_n\}_{n=0}^{\infty}$ and $\{b_n\}_{n=0}^{\infty}$ are periodic sequences with either the same period or with different periods. Eq. (6.5) consists of four cases. We will commence with the first special case of Eq. (6.5) in the **multiplicative form**:

$$x_{n+1} = a_n x_n, \quad n = 0, 1, \ldots, \tag{6.6}$$

where $x_0 \neq 0$ and $\{a_n\}_{n=0}^{\infty}$ is a period-k sequence for ($k \geq 2$). Our intent is to determine the periodic attributes of Eq. (6.6). The upcoming example will assume that $\{a_n\}_{n=0}^{\infty}$ as a period-2 sequence.

Example 6.1. *Solve Eq. (6.6) when $\{a_n\}_{n=0}^{\infty}$ is a period-2 sequence and determine its periodic traits.*

Solution: *By iterations and induction we get:*

$$x_0,$$
$$x_1 = a_0 x_0,$$
$$x_2 = a_1 x_1 = [a_1 a_0] x_0,$$
$$x_3 = a_0 x_2 = a_1 a_0^2 x_0,$$
$$x_4 = a_1 x_3 = [a_1 a_0]^2 x_0,$$
$$x_5 = a_0 x_4 = a_1^2 a_0^3 x_0,$$
$$x_6 = a_1 x_5 = [a_1 a_0]^3 x_0,$$
$$\vdots$$

Let $P = a_0 a_1$. Then for all $n \geq 0$:

$$\begin{cases} x_{2n} = P^n x_0, \\ x_{2n+1} = a_0 P^n x_0. \end{cases} \tag{6.7}$$

Thus, via Eq. (6.7), every solution of Eq. (6.6) is periodic with:

1. *Period-2 if P = 1.*

2. *Period-4 if P = −1.*

Now we will portray a graphical representation of a period-2 cycle when P = 1 and a period-4 cycle when P = −1. The consequent graph depicts a **positive period-2 cycle** *when $x_0 = 2$, $a_0 = 4$ and $a_1 = 0.25$:*

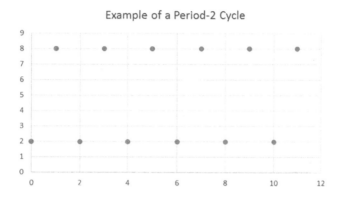

FIGURE 6.4   A positive period-2 cycle.

By designating distinct values of x_0, b_0 and b_1, we can acquire either a periodic cycle with all negative terms or a cycle with a positive and a negative term. The subsequent graph renders an **alternating period-4 cycle** *when $x_0 = 4$, $a_0 = 1$ and $a_1 = −1$:*

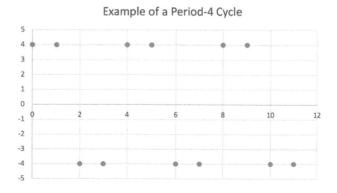

FIGURE 6.5   An alternating period-4 cycle.

Observe that Figure (6.4) depicts a positive period-2 cycle in comparison to Figure (6.5) that renders an alternating period-4 cycle as for all $n \geq 0$, $x_{n+2} = -x_n$. In fact if $P = -1$ then every solution of Eq. (6.6) is an alternating period-4 cycle with the corresponding pattern:

$$x_0, \ a_0 x_0, \ -x_0, \ -a_0 x_0, \ \ldots.$$

The next graph also depicts a period-4 cycle when $x_0 = 3$, $a_0 = 2$ and $a_1 = -0.5$:

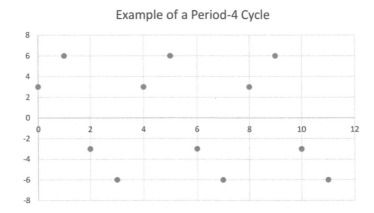

Example of a Period-4 Cycle

FIGURE 6.6   An alternating period-4 cycle.

*Observe that Figures (6.5) and (6.6) resemble a **mirror image** of the alternating terms. Figure (6.5) evokes a mirror image of steps as the two neighboring positive and negative terms as equal. On the contrary, Figure (6.6) describes a mirror image of the **ascending terms** and the **descending terms**; the cycle's positive ascending terms and the cycle's negative descending terms.*

The next example will outline the periodic features of Eq. (6.6) when $\{a_n\}_{n=0}^{\infty}$ is a period-3 sequence.

Example 6.2. *Solve Eq. (6.6) when $\{a_n\}_{n=0}^{\infty}$ is a period-3 sequence and determine its periodic traits.*

Solution: *Similar to Example (6.1) we acquire:*

$$x_0,$$
$$x_1 = a_0 x_0,$$
$$x_2 = a_1 x_1 = a_1 a_0 x_0,$$
$$x_3 = a_2 x_2 = [a_2 a_1 a_0] x_0,$$
$$x_4 = a_0 x_3 = a_2 a_1 a_0^2 x_0,$$
$$x_5 = a_1 x_4 = a_2 a_1^2 a_0^2 x_0,$$
$$x_6 = a_2 x_5 = [a_2 a_1 a_0]^2 x_0,$$
$$\vdots$$

Let $P = a_0 a_1 a_2$. Then for all $n \geq 0$:

$$\begin{cases} x_{3n} = P^n x_0, \\ x_{3n+1} = a_0 P^n x_0, \\ x_{3n+2} = a_0 a_1 P^n x_0. \end{cases} \tag{6.8}$$

Hence via Eq. (6.8), every solution of Eq. (6.6) is periodic with:

1. Period-3 if $P = 1$.

2. Period-6 if $P = -1$.

Now we will illustrate a sketch of a period-3 cycle when $P = 1$ and a period-6 cycle when $P = -1$ and compare their contrasts (a positive period-3 cycle and an alternating period-6 cycle). The next sketch depicts a "step-shaped" period-3 cycle when $x_0 = 2$, $a_0 = 2$, $a_1 = 1$ and $a_2 = 0.5$:

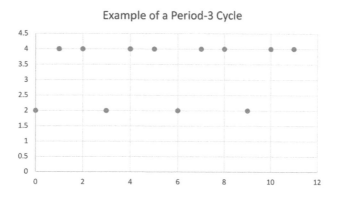

FIGURE 6.7   A step-shaped period-3 cycle.

Analogous to Example (6.1), we can select values of x_0, a_0, a_1 and a_2 and obtain either a negative period-3 cycle or a period-3 cycle with some terms that are positive and other terms that are negative. The consequent diagram traces a **decreasing period-3 cycle** *when $x_0 = 2$, $a_0 = -2$, $a_1 = 2$ and $a_2 = -0.25$:*

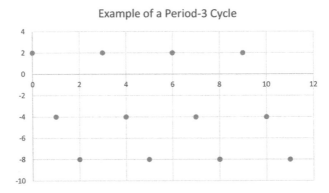

FIGURE 6.8   A period-3 cycle with one positive term and two negative terms.

First of all, notice that in Figure (6.7) all the terms of the periodic cycle are positive as x_0, a_0, a_1 and a_2 are positive. On the other hand, in Figure (6.8) one term of the periodic cycle is positive and two terms of the periodic cycle are negative. We can select related positive and negative values of x_0, a_0, a_1 and a_2 to acquire a periodic cycle with all negative terms. Additional distinct combinations of x_0, a_0, a_1 and a_2 can be selected to acquire specific positive and negative terms of the periodic cycle.

The next graph renders an **alternating period-6 cycle** *when $x_0 = -2$, $a_0 = 2$, $a_1 = 2$ and $a_2 = -0.25$:*

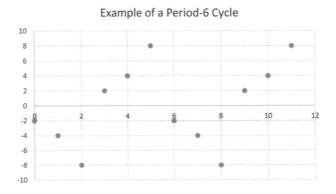

FIGURE 6.9   An alternating period-6 cycle.

Notice that Figure (6.9) renders an alternating period-6 cycle as for all $n \geq 0$, $x_{n+3} = -x_n$. In addition, if $P = -1$ then every solution of Eq. (6.6) is an alternating period-6 cycle.

Now suppose that $\{a_n\}_{n=0}^{\infty}$ is a period-4 sequence. The following graph describes a period-4 cycle with two positive terms and two negative terms when $x_0 = 2$, $a_0 = -2$, $a_1 = 0.5$, $a_2 = -4$ and $a_3 = 0.125$ (Figure 6.10):

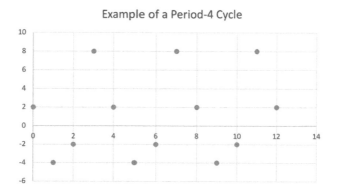

Example of a Period-4 Cycle

FIGURE 6.10   A period-4 cycle with two positive and two negative terms.

Analogous to Figures (6.5) and (6.9), the corresponding sketch renders an **alternating period-8 cycle** when $x_0 = 2$, $a_0 = 2$, $a_1 = 1$, $a_2 = 2$ and $a_3 = -0.25$ (Figure 6.11):

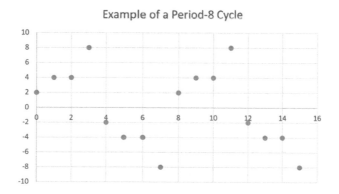

Example of a Period-8 Cycle

FIGURE 6.11   An alternating period-8 cycle.

Similar to Figures (6.5) and (6.9), we see that for all $n \geq 0$, $x_{n+4} = -x_n$.

From Examples (6.1) and (6.2) we will generalize the results. Suppose that $\{a_n\}_{n=0}^{\infty}$ is a period-k sequence ($k \geq 2$) and let $P = \prod_{i=0}^{k-1} a_i$. Then parallel to Examples (6.1) and (6.2), we obtain the following solution to Eq. (6.6):

$$
\begin{cases}
x_{kn} = P^n x_0, \\
x_{kn+1} = a_0 P^n x_0, \\
x_{kn+2} = a_0 a_1 P^n x_0, \\
\quad \vdots \\
x_{kn+k-2} = \left[\prod_{i=0}^{k-3} a_i \right] P^n x_0, \\
x_{kn+k-1} = \left[\prod_{i=0}^{k-2} a_i \right] P^n x_0.
\end{cases}
\tag{6.9}
$$

Hence via Eq. (6.9), every solution of Eq. (6.6) is periodic with:

1. Period-k if $P = 1$.

2. Period-$2k$ if $P = -1$.

Via Examples (6.1) and (6.2), we can conclude that when $P = -1$ then every solution of Eq. (6.6) is an **alternating period-$2k$ cycle**, where for all $n \geq 0$, $x_{n+k} = -x_n$. This will be left as an end-of-chapter exercise. Next we will examine the periodic character of the following special case of Eq. (6.5) in the **additive form**:

$$
x_{n+1} = x_n + b_n, \quad n = 0, 1, \ldots,
\tag{6.10}
$$

where $\{b_n\}_{n=0}^{\infty}$ is a period-k sequence ($k \geq 2$). Our goal is to determine the existence of periodic cycles of Eq. (6.10). The succeeding example will commence with $\{b_n\}_{n=0}^{\infty}$ as a period-2 sequence.

Example 6.3. *Solve Eq. (6.10) when $\{b_n\}_{n=0}^{\infty}$ is a period-2 sequence and determine its periodic traits.*

Solution: *Analogous to Example (6.1), by iterations we acquire:*

$$
\begin{aligned}
& x_0, \\
& x_1 = x_0 + b_0, \\
& x_2 = x_1 + b_1 = x_0 + [b_0 + b_1], \\
& x_3 = x_2 + b_0 = x_0 + b_0 + [b_0 + b_1], \\
& x_4 = x_3 + b_1 = x_0 + 2[b_0 + b_1], \\
& x_5 = x_4 + b_0 = x_0 + b_0 + 2[b_0 + b_1], \\
& x_6 = x_5 + b_1 = x_0 + 3[b_0 + b_1], \\
& \quad \vdots
\end{aligned}
$$

Let $S = b_0 + b_1$. Thus, for all $n \geq 0$:

$$\begin{cases} x_{2n} = x_0 + Sn, \\ x_{2n+1} = x_0 + b_0 + Sn. \end{cases} \tag{6.11}$$

The consequent graph sketches a **positive increasing period-2 cycle** *when $x_0 = 2$, $b_0 = 4$ and $b_1 = -4$ (Figure 6.12):*

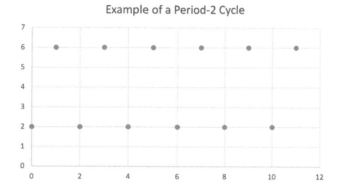

FIGURE 6.12   A positive increasing period-2 cycle.

Notice that via Eq. (6.11), every solution of Eq. (6.10) is periodic with period-2 if $S = 0$ with the following pattern:

$$x_0, \ x_0 + b_0, \ x_0, \ x_0 + b_0, \ \ldots .$$

By selecting different values of x_0, b_0 and b_1, we can obtain either a **negative periodic cycle** *or a periodic cycle with a positive and a negative term.*

Example 6.4. *Solve Eq. (6.10) when $\{b_n\}_{n=0}^{\infty}$ is a period-3 sequence and determine its periodic character.*

Solution: *Analogous to Example (6.3), by iterations we acquire:*

$$\begin{aligned}
&x_0, \\
&x_1 = x_0 + b_0, \\
&x_2 = x_1 + b_1 = x_0 + b_0 + b_1, \\
&x_3 = x_2 + b_2 = x_0 + [b_0 + b_1 + b_2], \\
&x_4 = x_3 + b_0 = x_0 + b_0 + [b_0 + b_1 + b_2], \\
&x_5 = x_4 + b_1 = x_0 + b_0 + b_1 + [b_0 + b_1 + b_2], \\
&x_6 = x_5 + b_2 = x_0 + 2[b_0 + b_1 + b_2]. \\
&\vdots
\end{aligned}$$

Let $S = b_0 + b_1 + b_2$. Hence, for all $n \geq 0$:

$$\begin{cases} x_{3n} = x_0 + Sn, \\ x_{3n+1} = x_0 + b_0 + Sn, \\ x_{3n+2} = x_0 + b_0 + b_1 + Sn. \end{cases} \qquad (6.12)$$

Via Eq. (6.12), every solution of Eq. (6.10) is periodic with period-3 if $S = 0$. Analogous to the previous examples, we can detect the structure's sensitivity of the periodic cycles depending on the values of x_0, b_0, b_1 and b_2. The corresponding graph renders an **increasing period-3 cycle** *when $x_0 = 1$, $b_0 = 3$, $b_1 = 1$ and $b_2 = -4$ (Figure 6.13):*

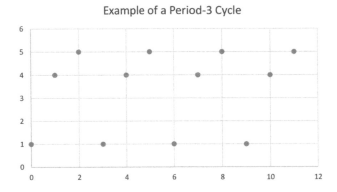

FIGURE 6.13   A positive period-3 cycle with increasing terms.

Switching the negative sign of b_0, b_1 or b_2 will affect the pattern of the periodic cycle. Analogous to Figures (6.8) and (6.6), we acquire an **increasing period-3 cycle** *with $x_0 = -3$, $b_0 = 2$, $b_1 = 4$ and $b_2 = -6$ (Figure 6.14):*

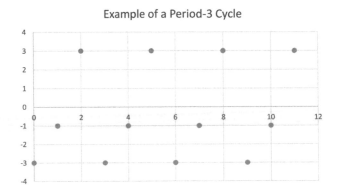

FIGURE 6.14   A period-3 cycle with two negative terms and a positive term.

Now suppose that $\{b_n\}_{n=0}^{\infty}$ is a period-4 sequence. The sketch below traces an **increasing period-4 cycle** with two negative and two positive terms when $x_0 = -2$, $b_0 = 1$, $b_1 = 2$, $b_2 = 3$ and $b_3 = -6$ (Figure 6.15):

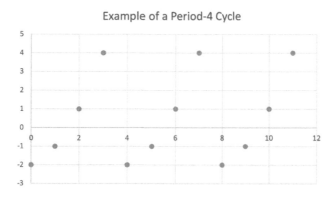

Example of a Period-4 Cycle

FIGURE 6.15   A period-4 cycle with two negative and two positive terms.

Now we will extend the results from Examples (6.3) and (6.4). Suppose that $\{b_n\}_{n=0}^{\infty}$ is a period-k sequence ($k \geq 2$) and let $S = \sum_{i=0}^{k-1} b_i$. Then we procure the following solution of Eq. (6.10):

$$
\begin{cases}
x_{kn} = x_0 + Sn, \\[4pt]
x_{kn+1} = x_0 + b_0 + Sn, \\[4pt]
x_{kn+2} = x_0 + b_0 + b_1 + Sn, \\[4pt]
\vdots \\[4pt]
x_{kn+k-2} = x_0 + \left[\sum_{i=0}^{k-3} b_i\right] + Sn, \\[4pt]
x_{kn+k-1} = x_0 + \left[\sum_{i=0}^{k-2} b_i\right] + Sn.
\end{cases}
\tag{6.13}
$$

Via Eq. (6.13), it follows that every solution of Eq. (6.10) is periodic with period-k if $S = 0$.

We will shift our focus on the periodic traits of the following special case of Eq. (6.5):

$$
x_{n+1} = -x_n + b_n, \quad n = 0, 1, \ldots,
\tag{6.14}
$$

where $\{b_n\}_{n=0}^{\infty}$ is a period-k sequence ($k \geq 2$). Our aims are to determine the existence and uniqueness of periodic cycles of Eq. (6.14). The succeeding example will emerge with $\{b_n\}_{n=0}^{\infty}$ as a period-2 sequence.

Example 6.5. *Suppose that $\{b_n\}_{n=0}^{\infty}$ is a period-2 sequence. Show that Eq. (6.14) has no period-2 cycles and explain why.*

Solution: *Suppose that $x_2 = x_0$. Then:*

$$x_0,$$
$$x_1 = -x_0 + b_0,$$
$$x_2 = -x_1 + b_1 = -[x_1 + b_0] + b_1 = x_0 - b_0 + b_1 = x_0.$$

Thus, $x_2 = x_0$ if and only if $b_0 = b_1$. This is a contradiction as we assumed that $\{b_n\}_{n=0}^{\infty}$ is a period-2 sequence where $b_0 \neq b_1$. The next sequence of examples will signify the contrasting periodic traits of Eq. (6.14) when $\{b_n\}_{n=0}^{\infty}$ is an even-ordered periodic sequence in comparison to when $\{b_n\}_{n=0}^{\infty}$ is an odd-ordered periodic sequence.

Example 6.6. *Suppose that $\{b_n\}_{n=0}^{\infty}$ is a period-3 sequence. Determine the necessary and sufficient conditions for the existence of period-3 cycles of Eq. (6.14).*

Solution: *By iterations we acquire:*

$$x_0,$$
$$x_1 = -x_0 + b_0,$$
$$x_2 = -[x_1] + b_1 = -[-x_0 + b_0] + b_1 = x_0 + b_1 - b_0,$$
$$x_3 = -[x_2] + b_2 = -[x_0 + b_1 - b_0] + b_2 = -x_0 + b_2 + b_0 - b_1 = x_0.$$

Hence,

$$x_0 = \frac{b_0 - b_1 + b_2}{2}, \tag{6.15}$$

*and we obtain the corresponding **unique** period-3 pattern:*

$$x_0 = \frac{b_0 - b_1 + b_2}{2}, \quad x_1 = \frac{b_1 - b_2 + b_0}{2}, \quad x_2 = \frac{b_2 - b_0 + b_1}{2}, \quad \ldots.$$

*This is the also first time that we encounter a **unique periodic cycle**. The consequent sketch renders a "**triangular-shaped**" period-3 cycle when $x_0 = 1$, $b_0 = 4$, $b_1 = 4$ and $b_2 = 2$ (Figure 6.16):*

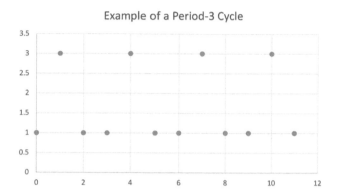

FIGURE 6.16   A triangular-shaped period-3 cycle.

Example (6.6) then leads us to the following result describing the unique periodic character of Eq. (6.14) when $\{b_n\}_{n=0}^{\infty}$ is an odd-ordered periodic sequence. Suppose that $\{b_n\}_{n=0}^{\infty}$ is a period-$(2k + 1)$ sequence, $(k \in \mathbb{N})$. Then Eq. (6.14) has a **unique periodic cycle** with period-$(2k+1)$ where

$$x_0 = \frac{\sum_{i=0}^{k+1} b_{2i} - \sum_{i=0}^{k} b_{2i+1}}{2} = \frac{\sum_{i=1}^{2k+1} (-1)^{i+1} b_{i-1}}{2}. \qquad (6.16)$$

First of all observe that Eq. (6.16) is an extension of Eq. (6.15). Second of all notice that via Eq. (6.15) we can conclude that Eq. (6.16) has $k + 1$ positive terms and k negative terms. In fact, the even-indexed coefficients are positive and the odd-indexed coefficients are negative. The proof of Eq. (6.16) will be left as an end-of-chapter exercise. Now we will direct our focus to when $\{b_n\}_{n=0}^{\infty}$ is an even-ordered periodic sequence.

Example 6.7. *Suppose that $\{b_n\}_{n=0}^{\infty}$ is a period-4 sequence. Determine the necessary and sufficient conditions for the existence of period-4 cycles of Eq. (6.14).*

Solution: *By iteration we procure:*

$$x_0,$$

$$x_1 = -x_0 + b_0,$$

$$x_2 = -[x_1] + b_1 = -[-x_0 + b_0] + b_1 = x_0 + b_1 - b_0,$$

$$x_3 = -[x_2] + b_2 = -[x_0 + b_1 - b_0] + b_2 = -x_0 + a_2 + a_0 - a_1$$

$$x_4 = -[x_3] + b_3 = -[-x_0 + b_2 + b_0 - b_1] + b_3,$$

$$= x_0 - [b_2 + b_0] + [b_1 + b_3] = x_0.$$

Period-4 cycles exist if and only if

$$b_1 + b_3 = b_0 + b_2 \tag{6.17}$$

with the corresponding period-4 pattern:

$$x_0, \ -x_0 + b_0, \ x_0 + b_1 - b_0, \ -x_0 + b_2 + b_0 - b_1, \ \dots.$$

*The ensuing graph renders a decreasing **step-shaped** period-4 cycle with two repeated positive terms, 0 and a negative term when $x_0 = 2$, $b_0 = 4$, $b_1 = 2$, $b_2 = -4$ and $b_3 = -2$ (Figure 6.17):*

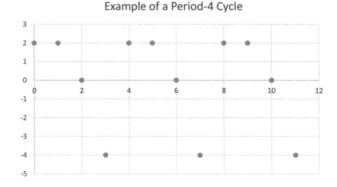

FIGURE 6.17 A decreasing **step-shaped** period-4 cycle.

Example (6.7) then guides us to the upcoming result that portrays the periodic traits of Eq. (6.14) when $\{b_n\}_{n=0}^{\infty}$ is an even-ordered periodic sequence. Suppose that $\{b_n\}_{n=0}^{\infty}$ is a period-$2k$ sequence, ($k \geq 2$). Then every solution of Eq. (6.14) is periodic with period-$2k$ if and only if:

$$\sum_{i=1}^{k} b_{2i-1} = \sum_{i=1}^{k} b_{2i-2}. \tag{6.18}$$

Notice that Eq. (6.18) extends directly from Eq. (6.17). In addition, the sum of all the odd-indexed coefficients must be equal to the sum of all the even-indexed coefficients. Furthermore, (6.18) has k terms on the left side and k terms on the right side. The proof of Eq. (6.18) will be left as an end-of-chapter exercise.

We will adjourn this chapter by examining the final case of Eq. (6.5) when $\{a_n\}_{n=0}^{\infty}$ and $\{b_n\}_{n=0}^{\infty}$ are periodic sequences. The next series of examples will assume that $\{a_n\}_{n=0}^{\infty}$ and $\{b_n\}_{n=0}^{\infty}$ are periodic with the same period. Eq. (6.5) will have unique periodic cycles in comparison to what we encountered in Example (6.6).

Example 6.8. *Suppose that $\{a_n\}_{n=0}^{\infty}$ and $\{b_n\}_{n=0}^{\infty}$ are period-2 sequences. Determine the pattern of the unique period-2 cycle of Eq. (6.5).*

Solution: *First set $x_2 = x_0$ and by iteration we get:*

$x_0,$
$x_1 = a_0 x_0 + b_0,$
$x_2 = a_1 [x_1] + b_1 = a_1 [a_0 x_0 + b_0] + b_1 = a_0 a_1 x_0 + a_1 b_0 + b_1 = x_0.$

We obtain

$$x_0 = \frac{a_1 b_0 + b_1}{1 - a_0 a_1} \tag{6.19}$$

provided that $a_0 a_1 \neq 1$ and the corresponding **unique period-2 pattern:**

$$\frac{a_1 b_0 + b_1}{1 - a_0 a_1}, \frac{a_0 b_1 + b_0}{1 - a_0 a_1}, \frac{a_1 b_0 + b_1}{1 - a_0 a_1}, \frac{a_0 b_1 + b_0}{1 - a_0 a_1}, \ldots \tag{6.20}$$

First of all we acquire a **unique period-2 cycle.** *Second of all, the indices of a_0, a_1, b_0 and b_1 in the numerator shift by an index of 1, while the terms in the denominator do not change.*

Now we will examine two graphical examples. The ensuing diagram renders (6.20) as an **increasing period-2 cycle** *with positive terms when* $x_0 = 3$, $a_0 = 1$, $a_1 = -1$, $b_0 = 2$ *and* $b_1 = 2$ *(Figure 6.18):*

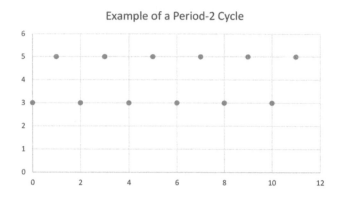

FIGURE 6.18   An increasing period-2 cycle with two positive terms.

We can select values of x_0, a_0, a_1, b_0 *and* b_1 *that will produce various combinations of period-2 cycles. For instance, all the terms of the cycle are negative, a cycle with one positive term and one negative term and a cycle with one 0 term. The next sketch depicts (6.20) as a negative* **decreasing period-2 cycle** *when* $x_0 = -1$, $a_0 = 1$, $a_1 = 3$, $b_0 = -1$ *and* $b_1 = 5$ *(Figure 6.19):*

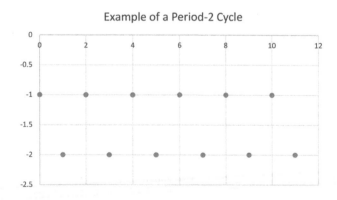

FIGURE 6.19   A decreasing period-2 cycle with two negative terms.

The succeeding example will decipher the periodic character of Eq. (6.5) when $\{a_n\}_{n=0}^{\infty}$ and $\{b_n\}_{n=0}^{\infty}$ are period-3 sequences.

Example 6.9. *Suppose that $\{a_n\}_{n=0}^{\infty}$ and $\{b_n\}_{n=0}^{\infty}$ are period-3 sequences. Determine the pattern of the unique period-3 cycle of Eq. (6.5).*

Solution: *Set $x_3 = x_0$ and by iteration we get:*

$$x_0,$$
$$x_1 = a_0 x_0 + b_0,$$
$$x_2 = a_1 [x_1] + b_1 = a_1 [a_0 x_0 + b_0] + b_1 = a_0 a_1 x_0 + a_1 b_0 + b_1,$$
$$x_3 = a_2 [x_2] + b_2 = a_2 [a_0 a_1 x_0 + a_1 b_0 + b_1] + b_2,$$
$$= a_0 a_1 a_2 x_0 + a_1 a_2 b_0 + a_2 b_1 + b_2 = x_0.$$

We acquire

$$x_0 = \frac{a_1 a_2 b_0 + a_2 b_1 + b_2}{1 - a_0 a_1 a_2} = \frac{a_2 [a_1 b_0 + b_1] + b_2}{1 - a_0 a_1 a_2}, \quad (6.21)$$

provided that $a_0 a_1 a_2 \neq 1$ and procure the corresponding unique period-3 pattern:

$$\frac{a_2 [a_1 b_0 + b_1] + b_2}{1 - a_0 a_1 a_2}, \frac{a_0 [a_2 b_1 + b_2] + b_0}{1 - a_0 a_1 a_2}, \frac{a_1 [a_0 b_2 + b_0] + b_1}{1 - a_0 a_1 a_2}, \dots. $$
$$(6.22)$$

Analogous to Example (6.8), the period-3 cycle in Eq. (6.22) is unique. In addition, the indices in the numerator shift by an index of 1 while the indices of the denominator do not change from term to term.

Now we will analyze three contrasting graphical examples of Eq. (6.22). The first sketch resembles a **step-shaped** *period-3 cycle with two positive terms and 0 when $x_0 = 1$, $a_0 = 3$, $a_1 = -1$, $a_2 = 1$, $b_0 = 2$, $b_1 = 4$ and $b_2 = 2$ (Figure 6.20):*

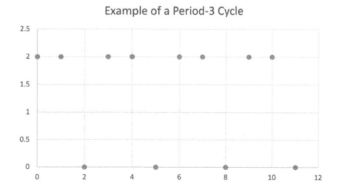

FIGURE 6.20   A step-shaped period-3 cycle with two positive terms and zero.

Analogous to Example (6.8), we can select additional values of x_0, a_0, a_1, b_0 and b_1 that will generate additional combinations of period-3 cycles. The succeeding diagram evokes a period-3 cycle with a positive term, negative term and zero when $x_0 = 4$, $a_0 = -2$, $a_1 = 0.5$, $a_2 = 1$, $b_0 = 2$, $b_1 = 3$ and $b_2 = 4$ (Figure 6.21):

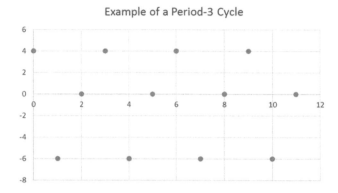

FIGURE 6.21   A period-3 cycle a positive term, negative term and zero.

The upcoming sketch renders a period-3 cycle when $x_0 = 1$, $a_0 = 3$, $a_1 = -1$, $a_2 = 1$, $b_0 = 2$, $b_1 = 4$ and $b_2 = 2$ (Figure 6.22):

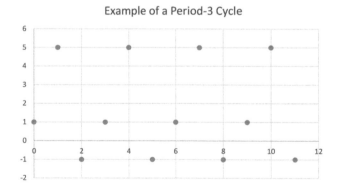

FIGURE 6.22   A period-3 cycle two positive terms and a negative term.

Examples (6.8) and (6.9) then lead us to the following result. Suppose that $\{a_n\}_{n=0}^{\infty}$ and $\{b_n\}_{n=0}^{\infty}$ are periodic sequences with period-k, $(k \geq 2)$, then Eq. (6.5) has a **unique period-k cycle** if

$$\prod_{i=0}^{k-1} a_i \neq 1,$$

and

$$x_0 = \frac{a_{k-1}\left[\ \ldots\ [a_3\,[a_2\,[a_1 b_0 + b_1] + b_2] + b_3]\ \ldots\ \right] + b_{k-1}}{1 - \left[\prod_{i=0}^{k-1} a_i\right]}. \qquad (6.23)$$

First of all (6.23) is expressed as a **Nested Sequence**. Alternatively, we can reformulate (6.23) as:

$$x_0 = \frac{\sum_{i=0}^{k-2}\left[\prod_{j=i+1}^{k-1} a_j b_i\right] + b_{k-1}}{1 - \left[\prod_{i=0}^{k-1} a_i\right]}. \qquad (6.24)$$

Second of all, the pattern of Eq. (6.23) emerges on the base of Eqs. (6.19) and (6.21). The proof of Eq. (6.23) or (6.24) is done by induction and will be left as an end-of-chapter exercise. Furthermore, the end-of-chapter exercises will examine the periodic character of Eq. (6.5) when the periodic sequences $\{a_n\}_{n=0}^{\infty}$ and $\{b_n\}_{n=0}^{\infty}$ are composed of different periods; for instance, when $\{a_n\}_{n=0}^{\infty}$ is periodic with period-2 and $\{b_n\}_{n=0}^{\infty}$ is periodic with period-3.

We will adjourn this chapter with the following special case of Eq. (6.5):

$$x_{n+1} = -a_n x_n + 1, \quad n = 0, 1, \ldots, \qquad (6.25)$$

where $\{a_n\}_{n=0}^{\infty}$ is a period-k sequence ($k \geq 2$). We will discover contrasting periodic patterns when $\{a_n\}_{n=0}^{\infty}$ is an even-ordered sequence in comparison to when $\{a_n\}_{n=0}^{\infty}$ is an odd-ordered periodic sequence. The subsequent examples will render these contrasting patterns.

Example 6.10. *Suppose that $\{a_n\}_{n=0}^{\infty}$ is period-2 sequence. Determine the pattern of the unique period-2 cycle of Eq. (6.25).*

Solution: *Set $x_2 = x_0$ and we acquire:*

$$x_0,$$
$$x_1 = -a_0 x_0 + 1,$$
$$x_2 = -a_1 x_1 + 1 = a_0 a_1 x_0 - a_1 + 1 = x_0.$$

We obtain

$$x_0 = \frac{-a_1 + 1}{1 - a_0 a_1} \qquad (6.26)$$

provided that $a_0a_1 \neq 1$ and acquire the corresponding unique period-2 pattern:

$$\frac{-a_1 + 1}{1 - a_0a_1}, \frac{-a_0 + 1}{1 - a_0a_1}, \dots \tag{6.27}$$

*Determining a formula for the unique period-2k cycle ($k \in \mathbb{N}$) when $\{a_n\}_{n=0}^{\infty}$ is period-2k sequence will be left as an end-of-chapter exercise. The subsequent sketch renders a **decreasing period-2 cycle** with a positive term and zero when $x_0 = 1$, $a_0 = 1$ and $a_1 = -4$ (Figure 6.23):*

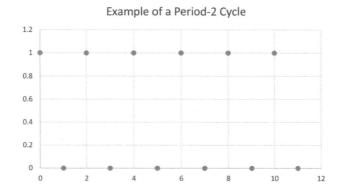

FIGURE 6.23   A decreasing period-2 cycle with a positive term and zero.

Example 6.11. *Suppose that $\{a_n\}_{n=0}^{\infty}$ is period-3 sequence. Determine the pattern of the unique period-3 cycle of Eq. (6.25).*

Solution: *Set $x_3 = x_0$ and we procure:*

$$
\begin{aligned}
x_0, &\\
x_1 &= -a_0x_0 + 1, \\
x_2 &= -a_1x_1 + 1 = a_0a_1x_0 - a_1 + 1, \\
x_3 &= -a_1x_2 + 1 = -a_0a_1a_2x_0 + a_1a_2 - a_2 + 1 = x_0.
\end{aligned}
$$

We acquire

$$x_0 = \frac{a_1a_2 - a_2 + 1}{1 + a_0a_1a_2} \tag{6.28}$$

provided that $a_0 a_1 a_2 \neq -1$ and the related unique period-3 pattern:

$$\frac{a_1 a_2 - a_2 + 1}{1 + a_0 a_1 a_2}, \quad \frac{a_2 a_0 - a_0 + 1}{1 + a_0 a_1 a_2}, \quad \frac{a_0 a_1 - a_1 + 1}{1 + a_0 a_1 a_2}, \ldots \quad (6.29)$$

*Obtaining a formula for the unique period-(2k+1) cycle ($k \in \mathbb{N}$) when $\{a_n\}_{n=0}^{\infty}$ is period-(2k+1) sequence will be left as an end-of-chapter exercise. The succeeding diagram traces a **triangular-shaped** period-3 cycle with three positive terms when $x_0 = 0.2$, $a_0 = 1$, $a_1 = 1$ and $a_2 = 4$ (Figure 6.24):*

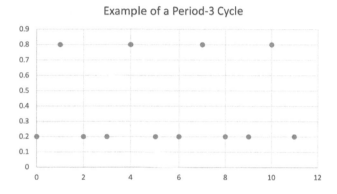

FIGURE 6.24   A triangular-shaped period-3 cycle with positive terms.

6.3 CHAPTER 6 EXERCISES

In problems 1–4, determine the periodic cycle by solving the given **Initial Value Problem**:

1:
$$\begin{cases} x_{n+1} = -x_n, & n = 0, 1, \ldots. \\ x_0 = -8. \end{cases}$$

2:
$$\begin{cases} x_{n+1} = -x_n, & n = 0, 1, \ldots. \\ x_0 = 6. \end{cases}$$

3:
$$\begin{cases} x_{n+1} = -x_n + 2, & n = 0, 1, \ldots. \\ x_0 = -6. \end{cases}$$

4:
$$\begin{cases} x_{n+1} = -x_n + 4, & n = 0, 1, \ldots. \\ x_0 = 4. \end{cases}$$

5:
$$\begin{cases} x_{n+1} = (-1)^{n+1} x_n, & n = 0, 1, \ldots. \\ x_0 = 2. \end{cases}$$

6:
$$\begin{cases} x_{n+1} = (-1)^n x_n, & n = 0, 1, \ldots. \\ x_0 = -1. \end{cases}$$

7:
$$\begin{cases} x_{n+1} = (-1)^n x_n + 1, & n = 0, 1, \ldots. \\ x_0 = 3. \end{cases}$$

8:
$$\begin{cases} x_{n+1} = (-1)^n x_n - 1, & n = 0, 1, \ldots. \\ x_0 = -2. \end{cases}$$

In problems 9–12, using Examples (6.1)–(6.2) and (6.9), solve the following recursive relation:

$$x_{n+1} = a_n x_n, \quad n = 0, 1, \ldots,$$

and determine the pattern of periodic cycles when:

9: $\{a_n\}_{n=0}^{\infty}$ is a period-2 sequence and $a_0 a_1 = -1$.

10: $\{a_n\}_{n=0}^{\infty}$ is a period-3 sequence and $a_0 a_1 a_2 = -1$.

11: $\{a_n\}_{n=0}^{\infty}$ is a period-4 sequence and $a_0 a_1 a_2 a_3 = -1$.

12: $\{a_n\}_{n=0}^{\infty}$ is a period-k sequence ($k \geq 2$) and $\prod_{i=0}^{k-1} a_i = -1$.

In problems 13–16, determine the periodic pattern of:

$$x_{n+1} = a_n x_n + 1, \quad n = 0, 1, \ldots,$$

13: when $\{a_n\}_{n=0}^{\infty}$ is a period-2 sequence (period-2 cycle).

14: when $\{a_n\}_{n=0}^{\infty}$ is a period-3 sequence (period-3 cycle).

15: when $\{a_n\}_{n=0}^{\infty}$ is a period-4 sequence (period-4 cycle).

16: when $\{a_n\}_{n=0}^{\infty}$ is a period-k sequence, ($k \geq 2$) (period-k cycle).

In problems 17–20, using Examples (6.10)–(6.11) determine the periodic pattern of:

$$x_{n+1} = -a_n x_n + 1, \quad n = 0, 1, \ldots,$$

17: when $\{a_n\}_{n=0}^{\infty}$ is a period-4 sequence (period-4 cycle).

18: when $\{a_n\}_{n=0}^{\infty}$ is a period-2k sequence, $(k \geq 1)$ (period-2k cycle).

19: when $\{a_n\}_{n=0}^{\infty}$ is a period-5 sequence (period-5 cycle).

20: when $\{a_n\}_{n=0}^{\infty}$ is a period-(2k+1) sequence, $(k \geq 1)$ (period-(2k+1) cycle).

In problems 21–24, using Example (6.6) determine the pattern of the unique periodic cycle of:

$$x_{n+1} = -x_n + b_n, \quad n = 0, 1, \ldots,$$

21: when $\{b_n\}_{n=0}^{\infty}$ is a period-5 sequence (period-5 cycle).

22: when $\{b_n\}_{n=0}^{\infty}$ is a period-7 sequence (period-7 cycle).

23: when $\{b_n\}_{n=0}^{\infty}$ is a period-9 sequence (period-9 cycle).

24: when $\{b_n\}_{n=0}^{\infty}$ is a period-(2k + 1) sequence, $(k \geq 1)$ (period-(2k + 1) cycle).

In problems 25–28, using Example (6.7) determine the pattern of periodic cycles of:
$$x_{n+1} = -x_n + b_n, \quad n = 0, 1, \ldots,$$

25: when $\{b_n\}_{n=0}^{\infty}$ is a period-6 sequence (period-6 cycles).

26: when $\{b_n\}_{n=0}^{\infty}$ is a period-8 sequence (period-8 cycles).

27: when $\{b_n\}_{n=0}^{\infty}$ is a period-10 sequence (period-10 cycles).

28: when $\{b_n\}_{n=0}^{\infty}$ is a period-2k sequence, $(k \geq 2)$ (period-2k cycles).

In problems 29–36, using Examples (6.8)–(6.9) determine the pattern of the unique periodic cycle of:

$$x_{n+1} = a_n x_n + b_n, \quad n = 0, 1, \ldots,$$

29: when $\{a_n\}_{n=0}^{\infty}$ and $\{b_n\}_{n=0}^{\infty}$ are period-4 sequences (period-4 cycle).

30: when $\{a_n\}_{n=0}^{\infty}$ and $\{b_n\}_{n=0}^{\infty}$ are period-5 sequences (period-5 cycle).

31: when $\{a_n\}_{n=0}^{\infty}$ and $\{b_n\}_{n=0}^{\infty}$ are period-6 sequences (period-6 cycle).

32: when $\{a_n\}_{n=0}^{\infty}$ and $\{b_n\}_{n=0}^{\infty}$ are period-k sequences, $(k \geq 2)$ (period-k cycle).

33: when $\{a_n\}_{n=0}^{\infty}$ is a period-2 sequence and $\{b_n\}_{n=0}^{\infty}$ is a period-4 sequence (period-4 cycle).

34: when $\{a_n\}_{n=0}^{\infty}$ is a period-4 sequence and $\{b_n\}_{n=0}^{\infty}$ is a period-2 sequence (period-4 cycle).

35: when $\{a_n\}_{n=0}^{\infty}$ is a period-3 sequence and $\{b_n\}_{n=0}^{\infty}$ is a period-6 sequence (period-6 cycle).

36: when $\{a_n\}_{n=0}^{\infty}$ is a period-6 sequence and $\{b_n\}_{n=0}^{\infty}$ is a period-3 sequence (period-6 cycle).

CHAPTER 7

Answers to Chapter Exercises

7.1 ANSWERS TO CHAPTER 1 EXERCISES

1. $\{3n\}_{n=1}^{\infty}$

3. $\{15n\}_{n=1}^{\infty}$

5. $\{4n\}_{n=5}^{\infty}$

7. $\{3n + 4\}_{n=0}^{\infty}$

9. $\{7n + 4\}_{n=0}^{\infty}$

11. $\{8 \cdot 2^n\}_{n=0}^{\infty} = \{2^{n+3}\}_{n=0}^{\infty}$

13. $\left\{\dfrac{3^n}{9}\right\}_{n=0}^{\infty} = \{3^{n-2}\}_{n=0}^{\infty}$

15. $\{5 \cdot 3^n\}_{n=0}^{\infty}$

17. $\left\{64 \cdot \left(\frac{3}{4}\right)^n\right\}_{n=0}^{\infty}$

19. For all $n \geq 0$:
$$x_n = \begin{cases} 4 & \text{if } n \text{ is even,} \\ -2 & \text{if } n \text{ is odd.} \end{cases}$$

21. For all $n \geq 0$:

$$x_n = \begin{cases} 1 & \text{if } n = 3k, \\ 3 & \text{if } n = 3k + 1, \\ -1 & \text{if } n = 3k + 2. \end{cases}$$

23. For all $n \geq 0$:

$$x_n = \begin{cases} n + 1 & \text{if } n \text{ is even,} \\ -[n + 1] & \text{if } n \text{ is odd.} \end{cases}$$

25. For all $n \geq 0$:

$$x_n = \begin{cases} 2^n & \text{if } n \text{ is even,} \\ -2^n & \text{if } n \text{ is odd.} \end{cases}$$

27. For all $n \geq 0$:

$$y = \begin{cases} -1 & \text{if } x \in (0, 2) \\ 1 & \text{if } x \in (2, 4) \\ 3 & \text{if } x \in (4, 6) \\ 5 & \text{if } x \in (6, 8) \\ \vdots \\ 2n - 1 & \text{if } x \in (2n, 2n + 2) \\ \vdots \end{cases}$$

29. For all $n \geq 0$:

$$y = \begin{cases} 2 - x & \text{if } x \in [0, 2] \\ x - 2 & \text{if } x \in [2, 4] \\ 6 - x & \text{if } x \in [4, 6] \\ x - 6 & \text{if } x \in [6, 8] \\ \vdots \\ (4n + 2) - x & \text{if } x \in [2n, 2n + 2] \\ x - (4n + 2) & \text{if } x \in [2n + 2, 2n + 4] \\ \vdots \end{cases}$$

31.

$$y = 0, \ x \in [0, 8],$$
$$y = 8 - x, \ x \in [0, 8],$$
$$x = 0, \ y \in [0, 8].$$

7.2 ANSWERS TO CHAPTER 2 EXERCISES

1. For all $n \geq 0$:

$$
y = \begin{cases}
2, \ x \in (0,1) \\
-4, \ x \in (1,3) \\
2, \ x \in (3,4) \\
-4, \ x \in (4,6) \\
\vdots \\
2, \quad x \in (3n, 3n+1) \\
-4, \quad x \in (3n+1, 3n+3) \\
\vdots
\end{cases}
$$

3. For all $n \geq 0$:

$$
y = \begin{cases}
1, \ x \in [0,1] \\
x - 0, \ x \in [1,3] \\
3, \ x \in [3,4] \\
x - 1, \ x \in [4,6] \\
\vdots \\
2n + 1, \quad x \in [3n, 3n+1] \\
x - n, \quad x \in [3n+1, 3n+3] \\
\vdots
\end{cases}
$$

5. For all $n \geq 0$:

$$
y = \begin{cases}
12, \ x \in [0,2] \\
14 - x, \ x \in [2,5] \\
9, \ x \in [5,7] \\
16 - x, \ x \in [7,10] \\
\vdots \\
12 - 3n, \quad x \in [5n, 5n+2] \\
(14 + 2n) - x, \quad x \in [5n+2, 5n+5] \\
\vdots
\end{cases}
$$

7. For all $n \geq 0$:

$$
y = \begin{cases}
x - 0, & x \in [0, 2] \\
4 - x, & x \in [2, 3] \\
x - 2, & x \in [3, 4] \\
6 - x, & x \in [4, 6] \\
\vdots \\
x - 6n, & x \in [6n, 6n + 2] \\
(4 + 6n) - x, & x \in [6n + 2, 6n + 3] \\
x - (2 + 6n), & x \in [6n + 3, 6n + 4] \\
(6 + 6n) - x, & x \in [6n + 4, 6n + 6] \\
\vdots
\end{cases}
$$

9.

$$
y = \begin{cases}
4, & x \in [-4, 4] \\
0, & x \in [0, 4] \\
-4, & x \in [-4, 4]
\end{cases}
\qquad
x = \begin{cases}
4, & y \in [-4, 4] \\
0, & y \in [-4, 4] \\
-4, & y \in [-4, 4]
\end{cases}
$$

11.

$$
y = \begin{cases}
4, & x \in [-4, 4] \\
0, & x \in [0, 4] \\
-4, & x \in [-4, 4] \\
-2, & x \in [2, 4]
\end{cases}
\qquad
x = \begin{cases}
4, & y \in [-4, 4] \\
0, & y \in [-4, 4] \\
-4, & y \in [-4, 4] \\
2, & y \in [-4, 0]
\end{cases}
$$

13.

$$
y = \begin{cases}
4, & x \in [-4, 4] \\
0, & x \in [0, 4] \\
-4, & x \in [-4, 4] \\
-2, & x \in [2, 4] \\
-3, & x \in [3, 4]
\end{cases}
\qquad
x = \begin{cases}
4, & y \in [-4, 4] \\
0, & y \in [-4, 4] \\
-4, & y \in [-4, 4] \\
2, & y \in [2, 4] \\
3, & y \in [-4, -2]
\end{cases}
$$

15.

$$y = \begin{cases} 4, & x \in [-4,4] \\ 0, & x \in [0,4] \\ -4, & x \in [-4,4] \\ -2, & x \in [2,4] \\ -3, & x \in [3,4] \\ -3.5, & x \in [3.5,4] \end{cases} \qquad x = \begin{cases} 4, & y \in [-4,4] \\ 0, & y \in [-4,4] \\ -4, & y \in [-4,4] \\ 2, & y \in [2,4] \\ 3, & y \in [-4,-2] \\ 3.5, & y \in [-4,-3] \end{cases}$$

17.

$$y = \begin{cases} 0, & x \in [0,3] \\ 1, & x \in [0,2] \\ 2, & x \in [0,1] \end{cases} \qquad x = \begin{cases} 0, & y \in [0,3] \\ 1, & y \in [0,2] \\ 2, & y \in [0,1] \end{cases} \qquad y = \begin{cases} 3-x, & x \in [0,3] \\ 2-x, & x \in [0,2] \\ 1-x, & x \in [0,1] \end{cases}$$

19.

$$y = \begin{cases} 0, & x \in [0,5] \\ 1, & x \in [0,4] \\ 2, & x \in [0,3] \\ 3, & x \in [0,2] \\ 4, & x \in [0,1] \end{cases} \qquad x = \begin{cases} 0, & y \subset [0,5] \\ 1, & y \in [0,4] \\ 2, & y \in [0,3] \\ 3, & y \in [0,2] \\ 4, & y \in [0,1] \end{cases} \qquad y - \begin{cases} 5-x, & x \in [0,4] \\ 4-x, & x \in [0,4] \\ 3-x, & x \in [0,3] \\ 2 & x, & x \in [0,2] \\ 1-x, & x \in [0,1] \end{cases}$$

21.

$$y = \begin{cases} 2, x \in [-2,2] \\ 4, x \in [-4,4] \\ 6, x \in [-6,6] \\ 8, x \in [-8,8] \end{cases} \qquad y = \begin{cases} x+0, x \in [0,8] \\ x+4, x \in [-2,4] \\ x+8, x \in [-4,0] \\ x+12, x \in [-6,-4] \end{cases} \qquad y = \begin{cases} 0-x, x \in [-8,0] \\ 4-x, x \in [-4,2] \\ 8-x, x \in [0,4] \\ 12-x, x \in [4,6] \end{cases}$$

23.

$$y = \begin{cases} 4, x \in [-4,4] \\ 6, x \in [-2,2] \\ 8, x \in [-8,8] \end{cases} \qquad y = \begin{cases} x+0, x \in [0,8] \\ x+4, x \in [0,2] \\ x+8, x \in [-4,0] \end{cases} \qquad y = \begin{cases} 0-x, x \in [-8,0] \\ 4-x, x \in [-2,0] \\ 8-x, x \in [0,4] \end{cases}$$

25.

$$y = \begin{cases} 4, x \in [-4, 4] \\ -4, x \in [-4, 4] \end{cases} \quad x = \begin{cases} 4, y \in [-4, 4] \\ -4, y \in [-4, 4] \end{cases} \quad y = \begin{cases} x - 4, x \in [0, 4] \\ x + 4, x \in [-4, 0] \\ 4 - x, x \in [0, 4] \\ -4 - x, x \in [-4, 0] \end{cases}$$

27.

$$y = \begin{cases} 4, x \in [-4, 4] \\ 2, x \in [-2, 2] \\ -2, x \in [-2, 2] \\ -4, x \in [-4, 4] \end{cases} \quad x = \begin{cases} 4, y \in [-4, 4] \\ 2, y \in [-2, 2] \\ -2, y \in [-2, 2] \\ -4, y \in [-4, 4] \end{cases}$$

$$y = \begin{cases} x - 4, x \in [0, 4] \\ x + 4, x \in [-4, 0] \\ x - 2, x \in [0, 2] \\ x + 2, x \in [-2, 0] \\ 2 - x, x \in [0, 2] \\ -2 - x, x \in [-2, 0] \\ 4 - x, x \in [0, 4] \\ -4 - x, x \in [-4, 0] \end{cases}$$

29.

$$y = \begin{cases} 0, x \in [0, 16] \\ 4, x \in [0, 4] \\ 8, x \in [0, 8] \end{cases} \quad x = \begin{cases} 0, y \in [0, 16] \\ 4, y \in [0, 4] \\ 8, y \in [0, 8] \end{cases} \quad y = \begin{cases} 4 - x, x \in [0, 4] \\ 8 - x, x \in [0, 8] \\ 16 - x, x \in [0, 16] \end{cases}$$

31.

$$y = \begin{cases} 0, x \in [0, 16] \\ 1, x \in [0, 1] \\ 2, x \in [0, 2] \\ 4, x \in [0, 4] \\ 8, x \in [0, 8] \end{cases} \quad x = \begin{cases} 0, y \in [0, 16] \\ 1, y \in [0, 1] \\ 2, y \in [0, 2] \\ 4, y \in [0, 4] \\ 8, y \in [0, 8] \end{cases} \quad y = \begin{cases} 1 - x, x \in [0, 1] \\ 2 - x, x \in [0, 2] \\ 4 - x, x \in [0, 4] \\ 8 - x, x \in [0, 8] \\ 16 - x, x \in [0, 16] \end{cases}$$

33.

$$
y = \begin{cases}
8, x \in [-8, 8] \\
6, x \in [-8, -6] \\
4, x \in [-8, -4] \\
2, x \in [-8, -2] \\
0, x \in [-8, 0] \\
-4, x \in [-8, 0] \\
-8, x \in [-8, 8]
\end{cases}
\qquad
x = \begin{cases}
8, y \in [-8, 8] \\
0, y \in [-8, 0] \\
-2, y \in [0, 2] \\
-4, y \in [-8, 4] \\
0, y \in [-8, 0] \\
-6, y \in [0, 6] \\
-8, y \in [-8, 8]
\end{cases}
$$

$$
y = \begin{cases}
x - 8, x \in [0, 4] \\
x - 0, x \in [-8, 0] \\
x + 4, x \in [-8, -2] \\
x + 8, x \in [-8, -4] \\
x + 12, x \in [-8, -6] \\
-x + 0, x \in [-8, -8] \\
-x \quad 4, x \in [-8, -2] \\
-x \quad 8, x \in [-8, 0] \\
-x - 12, x \in [-8, -6]
\end{cases}
$$

35. $\{3i + 2\}_{i=0}^{n}$

7.3 ANSWERS TO CHAPTER 3 EXERCISES

1. $\{4n + 1\}_{n=0}^{\infty}$

3. $\{9n - 1\}_{n=1}^{\infty}$

5. $\{(2n - 1)^2\}_{n=1}^{\infty}$

7. $\{(4n - 1)^2\}_{n=1}^{\infty}$

9. $\{(4n - 2)^2\}_{n=1}^{\infty}$

11. For all $n \geq 0$:

$$
\{x_n\}_{n=0}^{\infty} = \begin{cases}
4 & \text{if } n = 0, \\
4 + \left[\sum_{i=i}^{n} (2i + 1)\right] & \text{if } n \geq 1.
\end{cases}
$$

13. $\{2 \cdot 3^n\}_{n=0}^{\infty}$

15. $\left\{ 2 \cdot \left[\sqrt{3} \right]^n \right\}_{n=0}^{\infty}$

17. $\left\{ 12 \cdot \left(\frac{3}{2} \right)^n \right\}_{n=0}^{\infty}$

19. $\{x_n\}_{n=1}^{\infty} = \prod_{i=i}^{n} (2i - 1)$

21. For all $n \geq 0$:

$$\{x_n\}_{n=0}^{\infty} = \begin{cases} 1 & \text{if } n = 0, \\ \prod_{i=i}^{n} (4i - 1) & \text{if } n \geq 1. \end{cases}$$

23. For all $n \geq 0$:

$$\{x_n\}_{n=0}^{\infty} = \begin{cases} 5 & \text{if } n = 0, \\ 5 \cdot \left[\prod_{i=i}^{n} (i + 3) \right] & \text{if } n \geq 1. \end{cases}$$

25. For all $n \geq 0$:

$$\{x_n\}_{n=0}^{\infty} = \begin{cases} \frac{1}{8} & \text{if } n = 0, \\ \frac{1}{8} \cdot \left[\prod_{i=i}^{n} (2i) \right] & \text{if } n \geq 1. \end{cases}$$

27. For all $n \geq 0$:

$$\{x_n\}_{n=0}^{\infty} = \begin{cases} -[6n + 4] & \text{if } n \text{ is even,} \\ [6n + 4] & \text{if } n \text{ is odd.} \end{cases}$$

29. For all $n \geq 0$:

$$\{x_n\}_{n=0}^{\infty} = \begin{cases} (-1)^{\frac{n+2}{2}} 2^{\frac{n}{2}} & \text{if } n \text{ is even,} \\ \left[\sqrt{2} \right]^n & \text{if } n \text{ is odd.} \end{cases}$$

31. For all $n \geq 0$:

$$\{x_n\}_{n=0}^{\infty} = \begin{cases} (-1)^{\frac{n}{2}} 5[n + 1] & \text{if } n \text{ is even,} \\ (-1)^{\frac{n-1}{2}} 5[n + 1] & \text{if } n \text{ is odd.} \end{cases}$$

33. $3 \cdot 40 \cdot 81 = 9720$

35. $20 + 2 \cdot 20 \cdot 21 = 860$

37. $\dfrac{2^{10} - 1}{4}$

39. $\dfrac{3^9 + 1}{12}$

51. $\dfrac{[k - 6] \cdot [k + 7]}{2}$

53. $\dfrac{[k - 8] \cdot [k + 9]}{2}$

7.4 ANSWERS TO CHAPTER 4 EXERCISES

1. 84

3. 90

5. $\dbinom{7}{3}$

7. $[k + 2] \cdot [k + 1]$

9. $\prod_{i=1}^{n} [k + i]$

11. $8 \cdot 7 \cdot 6 \cdot 5 = \prod_{i=0}^{3} [8 - i]$

13. $\prod_{i=0}^{k-1} [2k - i]$

7.5 ANSWERS TO CHAPTER 5 EXERCISES

1. For all $n \geq 0$:
$$\begin{cases} x_{n+1} = x_n + 5, \\ x_0 = 2. \end{cases}$$

3. For all $n \geq 0$:
$$\begin{cases} x_{n+1} = x_n + 4, \\ x_0 = 5. \end{cases}$$

5. For all $n \geq 0$:
$$\begin{cases} x_{n+1} = x_n + 4(n + 1), \\ x_0 = 3. \end{cases}$$

7. For all $n \geq 0$:
$$\begin{cases} x_{n+1} = x_n + 3(2n + 1), \\ x_0 = 5. \end{cases}$$

9. For all $n \geq 0$:
$$\begin{cases} x_{n+1} = 3x_n, \\ x_0 = 4. \end{cases}$$

11. For all $n \geq 0$:
$$\begin{cases} x_{n+1} = \frac{2}{3}x_n, \\ x_0 = 54. \end{cases}$$

13. For all $n \geq 0$:
$$\begin{cases} x_{n+1} = 2(n+1)x_n, \\ x_0 = 1. \end{cases}$$

15. For all $n \geq 0$:
$$\begin{cases} x_{n+1} = (4n+5)x_n, \\ x_0 = 1. \end{cases}$$

17. For all $n \geq 0$:
$$\begin{cases} x_{n+1} = \frac{[2n+3]\cdot[2n+5]\cdot x_n}{2n+1}, \\ x_0 = 3. \end{cases}$$

19. For all $n \geq 0$:
$$\begin{cases} x_{n+1} = x_n + (2n+3), \\ x_0 = 1. \end{cases}$$

21. For all $n \geq 0$:
$$\begin{cases} x_{n+1} = x_n + (n+1)^2, \\ x_0 = 1. \end{cases}$$

29. For all $n \geq 0$, $x_n = \left(\frac{1}{2}\right)^{n+3}$.

31. For all $n \geq 0$, $x_n = \left(\frac{4}{3}\right)^{n+2}$.

33. For all $n \geq 0$, $x_n = \left(\frac{3}{7}\right)^n + 3$.

35. For all $n \geq 0$, $x_n = \frac{3-n}{2}$.

37. For all $n \geq 0$:
$$x_n = \begin{cases} -2 & \text{if } n \text{ is even,} \\ 9 & \text{if } n \text{ is odd.} \end{cases}$$

39. For all $n \geq 0$, $x_n = (-1)^n[n+4]$.

41. For all $n \geq 0$, $x_n = \sum_{i=0}^{n} 2^{i+2} = 4[2^{n+1} - 1]$.

43. For all $n \geq 0$, $x_n = \sum_{i=0}^{n} (i+1)^2 = \frac{n[n+1][2n+1]}{6}$.

7.6 ANSWERS TO CHAPTER 6 EXERCISES

1. For all $n \geq 0$:
$$x_n = \begin{cases} -8 & \text{if } n \text{ is even,} \\ 8 & \text{if } n \text{ is odd.} \end{cases}$$

3. For all $n \geq 0$:
$$x_n = \begin{cases} -6 & \text{if } n \text{ is even,} \\ 8 & \text{if } n \text{ is odd.} \end{cases}$$

5. For all $n \geq 0$:
$$x_n = \begin{cases} 2 & \text{if } n = 4k, \\ -2 & \text{if } n = 4k + 1, \\ -2 & \text{if } n = 4k + 2, \\ 2 & \text{if } n = 4k + 3. \end{cases}$$

7. For all $n \geq 0$:
$$x_n = \begin{cases} 3 & \text{if } n = 4k, \\ 4 & \text{if } n = 4k + 1, \\ -3 & \text{if } n = 4k + 2, \\ -2 & \text{if } n = 4k + 3. \end{cases}$$

9. Alternating period-4 pattern:
$$x_0, \ a_0 x_0, \ -x_0, \ a_0 x_0, \ \dots$$

11. Alternating period-8 pattern:
$$x_0, \ a_0 x_0, \ a_0 a_1 x_0, \ a_0 a_1 a_2 x_0, \ -x_0, \ -a_0 x_0, \ -a_0 a_1 x_0, \ -a_0 a_1 a_2 x_0, \ \dots$$

13. Unique period-2 pattern:
$$\frac{a_1 + 1}{1 - a_0 a_1}, \ \frac{a_0 + 1}{1 - a_0 a_1}, \ \dots$$

15. Unique period-4 pattern. For $i = 0, 1, 2, 3$:
$$\frac{a_{1+i} a_{2+i} a_{3+i} + a_{2+i} a_{3+i} + a_{3+i} + 1}{1 - a_0 a_1 a_2 a_3}.$$

17. Unique period-4 pattern. For $i = 0, 1, 2, 3$:
$$\frac{-a_{1+i} a_{2+i} a_{3+i} + a_{2+i} a_{3+i} - a_{3+i} + 1}{1 - a_0 a_1 a_2 a_3}.$$

19. Unique period-5 pattern. For $i = 0, 1, 2, 3, 4$:

$$\frac{a_{1+i}a_{2+i}a_{3+i}a_{4+i} - a_{2+i}a_{3+i}a_{4+i} + a_{3+i}a_{4+i} - a_{4+i} + 1}{1 + a_0 a_1 a_2 a_3 a_4}.$$

21. Unique period-5 pattern with:

$$x_0 = \frac{b_0 - b_1 + b_2 - b_3 + b_4}{2} = \frac{\sum_{i=0}^{2} b_{2i} - \sum_{i=0}^{1} b_{2i+1}}{2}.$$

23. Unique period-9 pattern with:

$$x_0 = \frac{\sum_{i=0}^{4} b_{2i} - \sum_{i=0}^{3} b_{2i+1}}{2}.$$

25. Provided that $\sum_{i=0}^{2} b_{2i} = \sum_{i=0}^{2} b_{2i+1}$, then for all $i \in \{0, 1, 2, 3, 4, 5\}$ we obtain a period-6 pattern:

$$x_i = (-1)^i x_0 + (-1)^{i+1} \sum_{j=0}^{i} (-1)^j b_j.$$

27. Provided that $\sum_{i=0}^{4} b_{2i} = \sum_{i=0}^{4} b_{2i+1}$, then for all $i \in \{0, 1, \ldots, 9\}$ we obtain a period-10 pattern:

$$x_i = (-1)^i x_0 + (-1)^{i+1} \sum_{j=0}^{i} (-1)^j b_j.$$

29. Unique period-4 pattern with:

$$x_0 = \frac{a_3 [a_2 [a_1 b_0 + b_1] + b_2] + b_3}{1 - a_0 a_1 a_2 a_3} = \frac{\sum_{i=0}^{2} \left[\prod_{j=i+1}^{3} a_j b_i \right] + b_3}{1 - a_0 a_1 a_2 a_3}.$$

31. Unique period-6 pattern with:

$$x_0 = \frac{a_5 [a_4 [a_3 [a_2 [a_1 b_0 + b_1] + b_2] + b_3] + b_4] + b_5}{1 - a_0 a_1 a_2 a_3 a_4 a_5}$$

$$= \frac{\sum_{i=0}^{4} \left[\prod_{j=i+1}^{5} a_j b_i \right] + b_5}{1 - a_0 a_1 a_2 a_3 a_4 a_5}.$$

33. Unique period-4 pattern with:

$$x_0 = \frac{a_1 \left[a_0 \left[a_1 b_0 + b_1\right] + b_2\right] + b_3}{1 - [a_0 a_1]^2} = \frac{\sum_{i=0}^{2} \left[\prod_{j=i+1}^{3} a_j b_i\right] + b_3}{1 - [a_0 a_1]^2}.$$

35. Unique period-6 pattern with:

$$x_0 = \frac{a_2 \left[a_1 \left[a_0 \left[a_2 \left[a_1 b_0 + b_1\right] + b_2\right] + b_3\right] + b_4\right] + b_5}{1 - [a_0 a_1 a_2]^2}$$

$$= \frac{\sum_{i=0}^{4} \left[\prod_{j=i+1}^{5} a_j b_i\right] + b_5}{1 - [a_0 a_1 a_2]^2}$$

Appendices

A.1 RIGHT TRIANGLES

1. **45−45−90 Triangles:**

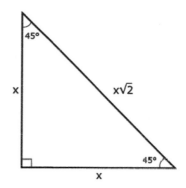

2. **30−60−90 Triangles:**

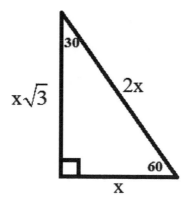

A.2 PATTERNS (SEQUENCES)

1. Linear Patterns:

$$1, 2, 3, 4, 5, 6, 7, \ldots = \{n\}_{n=1}^{\infty}$$

$$2, 4, 6, 8, 10, 12, 14, \ldots = \{2n\}_{n=1}^{\infty}$$

$$1, 3, 5, 7, 9, 11, 13, \ldots = \{2n+1\}_{n=0}^{\infty}$$

$$3, 6, 9, 12, 15, 18, 21, \ldots = \{3n\}_{n=1}^{\infty}$$

$$4, 8, 12, 16, 20, 24, 28, \ldots = \{4n\}_{n=1}^{\infty}$$

2. Quadratic Patterns:

$$1, 4, 9, 16, 25, 36, 49, \ldots = \{n^2\}_{n=1}^{\infty}$$

$$4, 16, 36, 64, 100, 144, 196, \ldots = \{(2n)^2\}_{n=1}^{\infty}$$

$$1, 9, 25, 49, 81, 121, 169, \ldots = \{(2n-1)^2\}_{n=1}^{\infty}$$

$$1, 25, 81, 169, 289, 441, 625, \ldots = \{(4n+1)^2\}_{n=0}^{\infty}$$

3. Geometric Patterns:

$$1, r, r^2, r^3, r^4, r^5, r^6, \ldots = \{r^n\}_{n=0}^{\infty}$$

$$1, 2, 4, 8, 16, 32, 64, \ldots = \{2^n\}_{n=0}^{\infty}$$

$$1, 3, 9, 27, 81, 243, 729, \ldots = \{3^n\}_{n=0}^{\infty}$$

$$1, 4, 16, 64, 256, 1024, 4096, \ldots = \{4^n\}_{n=0}^{\infty}$$

A.3 ALTERNATING PATTERNS (SEQUENCES)

1. Alternating Linear Patterns:

$$1, -2, 3, -4, 5, -6, 7, \ldots = \{(-1)^{n+1}\, n\}_{n=1}^{\infty}$$

$$-1, 2, -3, 4, -5, 6, -7, \ldots = \{(-1)^n\, n\}_{n=1}^{\infty}$$

$$1, -3, 5, -7, 9, -11, 13, \ldots = \{(-1)^n\, [2n+1]\}_{n=0}^{\infty}$$

$$-1, 3, -5, 7, -9, 11, -13, \ldots = \{(-1)^{n+1}\, [2n+1]\}_{n=0}^{\infty}$$

2. **Alternating Quadratic Patterns:**

$$1, \; -4, \; 9, \; -16, \; 25, \; -36, \; 49, \; \ldots = \{(-1)^{n+1} \, n^2\}_{n=1}^{\infty}$$

$$-1, \; 4, \; -9, \; 16, \; -25, \; 36, \; -49, \; \ldots = \{(-1)^{n} \, n^2\}_{n=1}^{\infty}$$

3. **Alternating Geometric Patterns:**

$$1, \; -r, \; r^2, \; -r^3, \; r^4, \; -r^5, \; r^6, \ldots = \{(-1)^{n} \, r^{n}\}_{n=0}^{\infty}$$

$$-1, \; r, \; -r^2, \; r^3, \; -r^4, \; r^5, \; -r^6, \ldots = \{(-1)^{n+1} \, r^{n}\}_{n=0}^{\infty}$$

A.4 PERIODIC PATTERNS AND MODULO ARITHMETIC

1. **Period-2 Pattern (Modulo 2):**

$$a_0, \; a_1, \; a_0, \; a_1, \; \ldots.$$

2. **Period-3 Pattern (Modulo 3):**

$$a_0, \; a_1, \; a_2, \; a_0, \; a_1, \; a_2, \; \ldots.$$

3. **Period-4 Pattern (Modulo 4):**

$$a_0, \; a_1, \; a_2, \; a_3, \; a_0, \; a_1, \; a_2, \; a_3, \; \ldots.$$

A.5 SUMMATION PROPERTIES

1. **Sigma Notation**

$$a_1 + a_2 + a_3 + a_4 + \ldots + a_n = \sum_{i=1}^{n} a_i.$$

2. **Addition of a Constant**

$$\sum_{i=1}^{n} c = c \cdot n.$$

3. **Distributive Property of Summations**

$$\sum_{i=1}^{n} [a_i \pm b_i] = \sum_{i=1}^{n} a_i \pm \sum_{i=1}^{n} b_i.$$

4. **Alternating Sums**

$$\sum_{i=1}^{n} (-1)^i a_i = -a_1 + a_2 - a_3 + a_4 - \ldots \pm a_n.$$

$$\sum_{i=1}^{n} (-1)^{i+1} a_i = a_1 - a_2 + a_3 - a_4 + \ldots \pm a_n.$$

A.6 FINITE SUMMATIONS

$1 + 2 + 3 + 4 + 5 + 6 + \ldots + n = \sum_{i=1}^{n} i = \frac{n[n+1]}{2}.$

$1 + 3 + 5 + 7 + 9 + 11 + \ldots + [2n - 1] = \sum_{i=1}^{n} (2i - 1) = n^2.$

$1 + 4 + 9 + 16 + 25 + 36 + \ldots + n^2 = \sum_{i=1}^{n} i^2 = \frac{n[n+1][2n+1]}{6}.$

$1 \cdot 2 + 2 \cdot 3 + 3 \cdot 4 + 4 \cdot 5 + \ldots + n \cdot [n + 1] = \sum_{i=1}^{n} i \cdot [i + 1] = \frac{n[n+1][n+2]}{3}.$

$\frac{1}{1 \cdot 2} + \frac{1}{2 \cdot 3} + \frac{1}{3 \cdot 4} + \frac{1}{4 \cdot 5} + \ldots + \frac{1}{n \cdot [n+1]} = \sum_{i=1}^{n} \frac{1}{i \cdot [i+1]} = \frac{n}{n+1}.$

$1 + r + r^2 + r^3 + r^4 + r^5 + \ldots + r^n = \sum_{i=0}^{n} r^i = \frac{1 - r^{n+1}}{1 - r}.$

$1 \cdot 2^0 + 2 \cdot 2^1 + 3 \cdot 2^2 + \ldots + n \cdot 2^{n-1} = \sum_{i=1}^{n} i \cdot 2^{i-1} = [n - 1]2^n + 1.$

$\binom{n}{0} + \binom{n}{1} + \binom{n}{2} + \ldots + \binom{n}{n-1} + \binom{n}{n} = \sum_{i=0}^{n} \binom{n}{i} = 2^n.$

Acknowledgments

$\mathbf{F}$IRST OF ALL, I would like to take the opportunity to thank the CRC Press editor and staff for their support, encouragement and guidance with this new textbook topic. Their encouragement certainly guided me in new innovative directions and to new experiences. Their suggestions were valuable with the textbook's structure such as necessary transitions between the sections, introduction of definitions and with the end-of-chapter exercises.

Second of all, I would like to thank the reviewers for their meticulous observations and suggestions while reviewing the book. Their vigilant comments were useful and enhanced the book's contents, transitions between topics and guided examples. Their prudent recommendations also directed me to new ideas for future textbooks.

In addition, I would like to thank the Rezekne Technical Academy High School director Aivars Vilkaste, Vineta Pavlova (English teacher) and the students for their support with the first pilot mini-course on "Introduction to Recognition of Patterns and Deciphering of Patterns" that I conducted there in May 2019. Their supportive feedback lead me to new valuable experiences and to new teaching innovations, practices and principles.

I would also like to thank my colleague Olga A. Orlova from the Munich Technical University for her artistic help with numerous diagrams. Olga emphasized several mistakes that she detected after meticulously checking each example in each section and in the end-of-chapter exercises. In addition, Olga suggested including specific examples of applications of geometric sequences and summations in geometry. In closing, Olga said, "This will be one of many books ahead."

Moreover, I would like to thank Daniil Timchenko, a high school student from Riga, Latvia, who identified several mistakes while he practiced working out the book's examples and the end-of-chapters exercises. Thanks to Daniil, the end-of-chapter exercises are more diverse and provide a broader range of challenge levels.

Finally, I would like to thank my parents Alexander and Shulamit for encouraging me to write textbooks, for their support with the textbook's content and for persuading me to continue writing future textbooks.

Bibliography

[1] A.M. Amleh, J. Hoag, G. Ladas, A Difference Equation with Eventually Periodic Solutions. *Computer & Mathematics with Applications*, 36 (1998), 401–404.

[2] A. Anisimova, M. Avotina, I. Bula, Periodic Orbits of Single Neuron Models with Internal Decay Rate $0 < \beta \leq 1$. *Mathematical Modelling and Analysis*, 18 (2013), 325–345.

[3] A. Anisimova, M. Avotina, I. Bula, Periodic and Chaotic Orbits of a Neuron Model. *Mathematical Modelling and Analysis*, 20 (2015), 30–52.

[4] W.F. Basener, B.P. Brooks, M.A. Radin, T. Wiandt, Rat Instigated Human Population Collapse on Easter Island. *Journal of Non-Linear Dynamics, Psychology and Life Sciences*, 12(3) (2008), 227–240.

[5] W.F. Basener, B.P. Brooks, M.A. Radin, T. Wiandt, Dynamics of a Population Model for Extinction and Sustainability in Ancient Civilizations. *Journal of Non-Linear Dynamics, Psychology and Life Sciences*, 12(1) (2008), 29–54.

[6] W.J. Briden, E.A. Grove, C.M. Kent, G. Ladas, Eventually Periodic Solutions of $x_{n+1} = max\{\frac{1}{x_n}, \frac{A_n}{x_{n-1}}\}$. *Communications on Applied Nonlinear Analysis*, 6(4) (1999).

[7] W.J. Briden, E.A. Grove, G. Ladas, L.C. McGrath, On the Non-autonomous Equation $x_{n+1} = max\{\frac{A_n}{x_n}, \frac{B_n}{x_{n-1}}\}$, *Proceedings of the Third International Conference on Difference Equations and Applications*. September 1–5, 1997, Taipei, Taiwan, Gordon and Breach Science Publishers (1999), 49–73.

[8] W.J. Briden, G. Ladas, T. Nesemann, On the Recursive Sequence $x_{n+1} = max\left\{\frac{1}{x_n}, \frac{A_n}{x_{n-1}}\right\}$. *Journal of Differential Equations and Applications*, 5 (1999), 491–494.

[9] I. Bula, M.A. Radin, N. Wilkins, Neuron Model with a Period Three Internal Decay Rate. *Electronic Journal of Qualitative Theory of Differential Equations* (EJQTDE), 46 (2017), 1–19.

[10] I. Bula, M.A. Radin, Periodic Orbits of a Neuron Model with Periodic Internal Decay Rate. *Applied Mathematics and Computation*, 266 (2015), 293–303.

[11] R.E. Crandall, On the '$3x + 1$' Problem. *Mathematics of Computation*, 32 (1978), 1281–1292.

[12] V. Da Fonte Dias, Signal Processing in the Sigma-Delta Domain. *Microelectronics Journal*, 26 (1995), 543–562.

[13] R.L. Devaney, A Piecewise Linear Model for the Zones of Instability of an Area-Preserving Map. *Physica D*, 10 (1984), 387–393.

[14] E.A. Grove, E. Lapierre, W. Tikjha, On the Global Behavior of $x_{n+1} = |x_n| - y_n - 1$ and $y_{n+1} = x_n + |y_n|$. *Cubo A Mathematica Journal*, 14(2) (2012), 111–152.

[15] C.M. Kent, E.A. Grove, G. Ladas, M.A. Radin, On $x_{n+1} = max\left\{\frac{1}{x_n}, \frac{A_n}{x_{n-1}}\right\}$ with a Period 3 Parameter, Fields Institute Communications, Volume 29, March 2001.

[16] C.M. Kent, M. Kustesky, A.Q. Nguyen, B.V. Nguyen, Eventually Periodic Solutions of $x_{n+1} = max\left\{\frac{A_n}{x_n}, \frac{B_n}{x_{n-1}}\right\}$, With Period Two Cycle Parameters.

[17] C.M. Kent, M.A. Radin, On the Boundedness of Positive Solutions of a Reciprocal Max–Type Difference Equation with Periodic Parameters. *International Journal of Difference Equations*, 8(2) (2013), 195–213.

[18] S.A. Kuruklis, G. Ladas, Oscillation and Global Attractivity in a Discrete Delay Logistic Model. *Applied Mathematics*, 50 (1992), 227–233.

[19] G. Ladas, On the Recursive Sequence $x_{n+1} = max\left\{\frac{A_0}{x_n}, \ldots, \frac{A_k}{x_{n-k}}\right\}$. *Journal of Differential Equations and Applications*, 2(2) (1996), 339–341.

[20] J.C. Lagarias, The $3x + 1$ Problem and Its Generalizations. *American Mathematical Monthly*, 92 (1985), 3–21.

[21] J.C. Lagarias, A. Weiss, The $3x + 1$ Problem: Two Stochastic Models. *Annals of Applied Probability*, 2 (1992), 229–261.

[22] E. Magnucka-Blandzi, J. Popenda, On the Asymptotic Behavior of a Rational System of Difference Equations. *Journal of Difference Equations and Applications*, 5(3) (1999), 271–286.

[23] O. Orlova, M. Radin, University Level Teaching Styles with High School Students and International Teaching and Learning, *International Scientific Conference "Society, Integration, Education"*, 2018.

[24] A.N. Pisarchik, M.A. Radin, R. Vogt, Non-autonomous Discrete Neuron Model with Multiple Periodic and Eventually Periodic Solutions. *Discrete Dynamics in Nature and Society*, Article ID 147282 (2015), 6 pages.

[25] M. Radin, *Periodic Character and Patterns of Recursive Sequences*, Springer, Cham, 2018.

[26] M. Radin, V. Riashchenko, Effective Pedagogical Management as a Road to Successful International Teaching and Learning. *Forum Scientiae Oeconomia*, 5(4) (2017), 71–84.

[27] N.F. Rulkov, Modeling of Spiking-Bursting Neural Behavior Using Two-Dimensional Map. *Physical Review E*, 65 (2002), 041922.

[28] R.J. Sacker, A Note on Periodic Ricker Maps. *Journal of Difference Equations and Applications*, 13(1) (2007), 89–92.

[29] P.A. Samuelson, Interaction between the Multiplier Analysis and the Principle of Acceleration. *The Review of Economics and Statistics*, 21 (1939), 75–78.

[30] J. Steele, Human Dispersals: Mathematical Models and the Archeological Record. *Human Biology*, 81(2/3) (2009), 121–140.

[31] Z. Yuan, L. Huang, All Solutions of a Class of Discrete-Time Systems Are Eventually Periodic. *Applied Mathematics and Computation*, 158 (2004), 537–546.

Index

Note: Page numbers in italics refer to figures.